Predicting the Unknown

The History and Future of Data Science and Artificial Intelligence

Stylianos Kampakis, PhD, CStat

Apress®

Predicting the Unknown: The History and Future of Data Science and Artificial Intelligence

Stylianos Kampakis
London, UK

ISBN-13 (pbk): 978-1-4842-9504-5   ISBN-13 (electronic): 978-1-4842-9505-2
https://doi.org/10.1007/978-1-4842-9505-2

Managing Director, Apress Media LLC: Welmoed Spahr
Acquisitions Editor: Shiva Ramachandran
Development Editor: James Markham
Editorial Assistant: Shaul Elson
Copy Editor: Kezia Endsley

Cover designed by eStudioCalamar

Distributed to the book trade worldwide by Springer Science+Business Media New York, 1 New York Plaza, Suite 4600, New York, NY 10004-1562, USA. Phone 1-800-SPRINGER, fax (201) 348-4505, e-mail orders-ny@springer-sbm.com, or visit www.springeronline.com. Apress Media, LLC is a California LLC and the sole member (owner) is Springer Science + Business Media Finance Inc (SSBM Finance Inc). SSBM Finance Inc is a **Delaware** corporation.

For information on translations, please e-mail booktranslations@springernature.com; for reprint, paperback, or audio rights, please e-mail bookpermissions@springernature.com.

Apress titles may be purchased in bulk for academic, corporate, or promotional use. eBook versions and licenses are also available for most titles. For more information, reference our Print and eBook Bulk Sales web page at http://www.apress.com/bulk-sales.

Any source code or other supplementary material referenced by the author in this book is available to readers on GitHub via the book's product page, located at www.apress.com/. For more detailed information, please visit http://www.apress.com/source-code.

Printed on acid-free paper

To all those who face uncertainty and overcome it.

To every person who came to be, is, and will be. Because uncertainty is a fundamental part of the human condition.

To all those who have helped reduce uncertainty throughout my life. I sincerely thank you.

Table of Contents

About the Author

Dr. Stylianos (Stelios) Kampakis is a data scientist, data science educator, and blockchain expert with more than ten years of experience. He has worked with decision makers from companies of all sizes: from startups to organizations like the U.S. Navy, Vodafone, and British Land. His work spans multiple sectors, including fintech (fraud detection and valuation models), sports analytics, health tech, general AI, medical statistics, predictive maintenance, and others. He has worked with many technologies, from statistical models to deep learning to blockchain. He also has two patents pending and has helped many people begin a career in data science and technology.

Stylianos is a member of the Royal Statistical Society, an honorary research fellow at the UCL Centre for Blockchain Technologies, a data science advisor for the London Business School, CEO of the Tesseract Academy, and a Tokenomics Auditor at Hacken. As a well-known data science educator, he has published two books, both with five-star ratings on Amazon. His personal website gets more than 10,000 visitors per month, and he is also a data science influencer on LinkedIn.

Acknowledgments

There are many, many individuals who influenced the development and writing of this book, directly or indirectly. I want to thank Lorena Goldsmith for her great editing and proofreading skills. She helped shape the narrative and readability of this book.

I also want to thank Morwenna Loughman for her editing, which greatly helped spice up the (admittedly a bit dry) academic writing style and turn it into a compelling story.

I also want to thank my former supervisor at the University College London, Professor Philip Treleaven, for helping me shape my writing. I want to thank my former thesis supervisor, Professor Philip Kargopoulos, for igniting my interest in some of the topics discussed in this book, such as the theory of the mind. Without him, the inspiration behind this book wouldn't exist.

Finally, I want to thank Dr. Theodosis Mourouzis, for his ideas in different areas of this book, especially regarding blockchain.

Preface

Author's Note to the Curious Reader

Uncertainty takes us on a journey through ancient philosophy, to renaissance mathematicians, to modern machine learning practitioners. It discusses how our minds and our brains handle uncertainty, and how and why the uncertainty caused by humans might be unconquerable. It also discusses modern technological developments, such as machine learning and blockchain, and explains how they affect the economy and the world around us.

I make the case that the common thread that connects the history of humankind to our current technological progress is uncertainty.

In the first chapter we'll explore a brief history of uncertainty – how the unknown has shaped civilisation, the manifold ways that it can encroach on our lives, and the different kinds and guises of uncertainty.

In the following chapter, we'll tackle one of the most important philosophical problems. Inductive reasoning is the process through which we create theories about the world; it is the foundation of scientific thinking. Methodologically speaking, it is one of the most important tools in our arsenal in the war on uncertainty. But why and how it works are questions that have baffled thinkers for more than two thousand years. We'll discuss how the problem was approached by philosophers of different eras, and what this has to do with some of the current tools that we use against uncertainty.

In chapters 3 to 5 we'll take a look at probability theory. We will see how this theory came to be, and we'll address some alternative theories to probability.

In chapter 6, we'll take a look at information theory, and the work of Claude Shannon – the first person to quantify uncertainty.

In chapter 7 we'll dive into the first discipline that explicitly dealt with the subject of uncertainty: statistics. The discipline of statistics monopolised the conversation about uncertainty until our recent developments in the computer era. We'll be looking at some of the core tools that statisticians are using, and some of the heroes – and controversial figures – of this field, such as Ronald Fisher, Karl Pearson and Egon Pearson.

Chapter 8 will explore the most current technology we have to describe, control and stave off uncertainty: machine learning. In the last two to three decades, machine learning has become the de facto way of attacking problems relating to prediction. We'll take a look at the basic types of machine learning, and discuss a few things about its history, the difference to statistics and some theories as to why it works so well.

In chapter 9 we'll be crossing what might be the next frontier in the study of uncertainty: causality, and how statistics and machine learning are falling short of understanding it. We'll address how – and why – the key to building true artificial intelligence might lie in causal relationships.

In chapter 10 we'll explore arguably one of the most exciting – and certainly the most contentious and risky – areas of uncertainty: forecasting. Predicting the future is something that has always captured the human imagination. From ancient oracles to pollsters, our fascination with what happens next might even be the most persuasive difference between humans and animals.

In chapters 11 and 12 we will deliberate the limits of prediction. Has nature imposed some kind of ceiling to hinder our ability to make accurate predictions and lift the veil of uncertainty? There have been many different theories across the centuries.

In chapter 13 we'll explore the role that uncertainty has played in the development of our brains and psychology. This has consequences in our everyday lives, which then ripple across society. We'll look at some of the heroes of this field, like Kahneman and Tversky, and the latest theories around uncertainty in the brain – such as Karl Friston's free energy principle.

In chapter 14 we will talk about a new kind of technology, one which promises to eradicate uncertainty in transactions. Called the "internet of trust", blockchain has promised to revolutionise many industries, from finance to supply chain. We'll discuss how it works, and whether those promises hold any merit.

Finally, in chapter 15 we'll cover the impact that uncertainty has on the global economy. We will break down industries into different categories based on how uncertainty influences them, and we'll discuss how current and future developments are going to change the face of those industries – and whether we can really talk about a new industrial revolution.

It's not easy to treat a complicated subject like uncertainty within the confines of a single book. A full treatment of uncertainty has to take into account many fields, from philosophy, to mathematics, to computer science, to biology to economics. Each of those fields has its own language, theories, heroes, and methods, and each warrants a whole tome of its own.

My goal in this book is to offer insight into the most important (in my opinion) parts from every field and approach. My ambition is for the reader to understand how uncertainty has shaped our lives, and to appreciate the tools and methods we might use against it. If you feel even just a little wiser and more enlightened after reading this book, I know that I accomplished my mission.

I am a data scientist, and I have spent most of my life trying to measure and control uncertainty. I've had the good fortune to do this both as a practitioner and as an academic. This gives me a unique perspective of the current technologies and algorithms being deployed to fight uncertainty, including their shortcomings. While the progress we've made in intelligent algorithms is noteworthy, we often forget our historical roots and the wider context. This book exists to link many historical trends, topics, and contexts under a common thread: the fight against uncertainty.

I hope that you enjoy reading this book as much as I enjoyed writing it. I cannot be certain that this is the right book for you, but what I *am* certain of is that after reading it, you will never look at the unknown, the unpredictable, and the uncertain in quite the same way again.

Prologue

A man in a tribe in 8,000 BC is delighted to have killed a gazelle. However, this happiness is short lived, as the food is barely enough for a family of five. Will he be able to find food the next day? A high school graduate just received a letter from Princeton. She holds up the letter with a mixed sense of anxiety and thrill and gets ready to open it. A whole nation is mesmerized by watching the returns. Which party is going to rule the country for the next four years? A couple is on their way to the airport to go on their scheduled holiday when they learn that all flights have been canceled due to a pandemic.

Experts debate whether humanity is facing environmental catastrophe in the 21st century. As the debate rages on, different opinions arise about the impact and the possible solutions to follow, but no one is sure what will happen.

What do all these, seemingly unrelated, anecdotes have in common?

Uncertainty is a common thread of existence. Recent advances in technology are simply the materialization of efforts to fight this enemy that humanity has faced since its beginnings. Philosophers dealt with questions around the nature of true knowledge and the state of the world. Mathematicians came up with ways to measure uncertainty and its opposite—information. Uncertainty is a part of life, one many of us would rather not live with, but one that we can't avoid.

The world is changing, and one of the major changes we're seeing is the increase of information, interconnectedness, measurement, control, and prediction. In our 24/7 connected world, there are few secrets left. We do not believe that dragons live at the edge of the world, or that monsters exist in forests outside our cities. Rather, those dragons and monsters now live in the future: in the unpredictable financial crisis, the unexpected illness, the unanticipated traffic jam that might cost us our job, or the unforeseen consequences of our choices, be they political or personal. We are gradually conquering the frontier of uncertainty, but some walls still seem impenetrable.

CHAPTER 1

Where Are We Now? A Brief History of Uncertainty

"Knowledge would be fatal. It is the uncertainty that charms one. A mist makes things wonderful."

—Oscar Wilde, *The Picture of Dorian Gray*

Think about the things in life of which you're certain. It could be things that you know about yourself: your name, your hair color, your address. There are also things that you likely know about your immediate physical environment: what the building next door looks like, what color the tree outside your window is today, or the name of your nearest supermarket. Without wading into an epistemological quagmire of how exactly we qualify these "knowns," they are, well, *known*.

You might also be fairly certain what the weather will be tomorrow, or what the inflation rate will be for the next three months. But how certain can you be about those beliefs? What if I asked you how certain you feel about the inflation rate in 12 months' time? Or the temperature seven years from now? You might be able to give me a plausible range, but could you give me the exact temperature of a specific location down to the tenth decimal digit?

Humanity has long fought a war against uncertainty. The powers of logic, science, and reason have enabled humans to create tools, methods, and, more recently, algorithms that can help us predict the future and control uncertainty. The hope, of course, is that in mitigating the unknown, we can improve our lives.

© Stylianos Kampakis 2023
S. Kampakis, *Predicting the Unknown*, https://doi.org/10.1007/978-1-4842-9505-2_1

If we look back a few thousand—or even just a couple of hundred—years, our ability to eliminate uncertainty has grown substantially. Our ancestors had rudimentary ways of predicting weather and agricultural patterns; they lacked knowledge of their wider social environments. They did not know whether they would be attacked and killed by some unseen neighboring tribe, or whether an unknown disease would wipe them out. Consider the advent of the telephone and imagine telling someone in 1877 that one day this new-fangled communication device would be able to tell them who was calling, before they even answered!

Technology has given us the power to make predictions about the future and the current state of the world, but how far can it take us? Is it limitless? Or are there some bounds that we can never surpass? And—crucially—have we caused more uncertainty in the process?

One of the most pressing questions—and one that it is my aim to explore in these pages—is how our current methods of prediction affect social dynamics, the economy, and the generalized uncertainty that we face. We might not always recognize the impact of uncertainty on our daily lives. But the role it plays should not be underestimated; of that we can be certain.

Over the last centuries, religion gave way to science, which promised—and to an extent delivered—explanations of the physical world and answers to the previously ineffable. The success of the scientific methods up until the 19th and 20th centuries gave rise to an optimism about the potential of science to explain everything by reducing every phenomenon down to its constituent parts. But in 1927, quantum mechanics and Heisenberg's uncertainty principle set some boundaries around what science could tell us—the limit does exist.

The discipline of statistics was created in the 19th century to deal with some of the unknowns that scientific disciplines faced. Ronald Fisher, though a controversial figure and a proponent of eugenics, was one of the fathers of modern statistics. He devised ways to test the validity of the insights gleaned from agricultural experiments. The same methods were then applied across other arenas, from physics, to evolution, to economics.

But the influence of uncertainty is not limited to the external world. It is very much a part of our internal world and functioning. Our brains have evolved partly as a mechanism to deal with uncertainty. Even simple single-cell organisms can learn to respond to series of stimuli,[1] effectively trying to predict what the next stimulus will look like and what the correct response will be.

This has shaped our minds. One of the current theories about the workings of the mind, which I discuss later in more depth, is Karl Friston's free energy principle— essentially that the prime goal of the brain is to reduce uncertainty. This has influenced the development of societies and most (or even all) societal structures. The legal system will punish someone who committed a crime, taking into account factors like their chance of recommitting the crime. That's not something we know, but the uncertainty around it influences our decision. The merchant believes that the economy tomorrow will be as good as, or at least not much worse than, it is today. The merchant can't know for sure. But this risk can't be avoided. The entrepreneur starts a new business believing that it will eventually take off. But the unknowns faced along the journey will make or break that business.

Humanity is currently facing some profound challenges. We could argue that the emergence of a hyper-connected globalized world partly evolved as a way to fend off the uncertainties of life. However, the very structure that has protected us for so long has also been the cause of problems at an unprecedented scale. Global warming, nuclear threat, another financial crisis, political upheaval around the world, or some kind of super-intelligent AI taking over—all of these are some of the uncertainties that we and future generations will have to deal with. They are also to a great extent self-inflicted— indeed humanity is one capricious mess of unpredictability. Can anyone guarantee how future events will unfold? Can anyone guarantee the absence or presence of disasters of a global magnitude? And does anyone really know the best response to those events? Our survival as a species up to this point has been partly due to the tools and methods we developed to defend against uncertainty—mostly against the hostile nature of disease and natural disasters. Our survival from now on will depend on how well we do with the uncertainty of events largely caused by us.

---

[1] www.sciencedaily.com/releases/2016/04/160427081533.htm

Not All Uncertainty Is Created Equal

A good way to start talking about uncertainty is to address the heterogeneous kinds of uncertainty. The first thing we must understand is that uncertainty can come in different shapes and sizes. For example:

1) We face uncertainty about the future. For example, what will the weather be like tomorrow? Will the economy be in good shape in two years?

2) We face uncertainty about right now. For example, the Secret Service might be wondering whether there is someone right now planning an attack.

3) We face uncertainty about the past. Think about all the historical events for which we have no clear idea why or what really happened. What about the Kennedy assassination? Or history before the written word?

4) We face uncertainty about measurements and data collection. How can we be 100 percent certain that, when we measure a specific quantity, our measurement is correct? Sometimes instruments can be faulty. Maybe a blood sample is corrupted. Maybe we conducted a poll, and the data collection process was biased.

5) We face uncertainty around cause and effect. How can we know that a medicine is really effective? How can an intervention in an economy change the inflation rate? How do we know that a policy is effective?

6) We face uncertainty around our choices. Faced with multiple decisions, how can we know which one is the best? Think about every decision you've made in your life, from whether or not to go to college, to whether or not to go out on Saturday night. How do these decisions influence your life? Even the seemingly inconsequential and flippant calls can have a lasting impact.

These are only some of the types of uncertainty that we face on a regular basis. Let's take a closer look at one of the most common categorizations of uncertainty. I use two examples and asses how these categorizations relate to the examples:

- ***Example #1:*** You are a hedge fund manager and are thinking of investing in a tech company. You assess their product and IP. After a few days of analysis, you judge this to be a good proposition. You invest $1million. The investment pays off, and within two years you have a 30 percent return.

- ***Example #2:*** You are a hedge fund manager and are thinking of investing in a tech company. After many months of analysis and due diligence, you decide to invest in the company. They have a solid proposition. You invest $1 million, but the next day the CEO suffers a fatal and completely unexpected heart attack. The company fails, and you lose your investment.

You might be thinking that the second example is far-fetched. How often do things like that *actually* happen? Well, how about a CEO suddenly dying, taking $190 million to their grave?

About $190m in cryptocurrency has been locked away in an online black hole after the founder of a currency exchange died, apparently taking his encrypted access to their money with him.

Investors in QuadrigaCX, Canada's largest cryptocurrency exchange, were unable to access their funds after its founder, Gerald Cotten, died on a trip to India in 2018. He was 30 years old.

According to a court filing first reported by CoinDesk, a cryptocurrency news and events company, Jennifer Robertson—identified as Cotten's widow—said the exchange owes its customers roughly C$250m (US$190m) in cash and cryptocurrency held in its "cold storage."

These examples illustrate the two types of uncertainty: aleatoric uncertainty and epistemic uncertainty (bear with me):

- **Epistemic uncertainty** refers to the kind of uncertainty that exists due to a lack of data or appropriate methods. If we only had more data or better methods of data collection, epistemic uncertainty would disappear.

- **Aleatoric uncertainty** also goes by the name of "statistical uncertainty." It refers to uncertainty inherent in the system. This is uncertainty we can't eliminate, but we can control it using tools from statistics or machine learning.

In example #1, the hedge fund manager spent only a few days analyzing the company, whereas in the second case, they spent months. In theory, someone could argue that the manager in both examples faced only epistemic uncertainty. Given all possible knowledge about the company, the CEO of the company, and every other possible factor, would make it possible to predict a positive outcome—in other words, the 30 percent return.

But there are two arguments here that turn this into a case of aleatoric uncertainty. The first is that, given the constraints that the manager operated under, knowing that someone suffered from a serious medical condition might have been impossible. There was really no way to possess this knowledge under any circumstances.

But let's say that actually it *would* be possible to get this kind of knowledge. Maybe hiring a detective would do the trick. However, even if we knew that this person suffers from a condition, we could only know a probability of dying. We can't be 100 percent certain about whether this person will die from the disease and when.

This example also demonstrates that the boundaries between aleatoric and epistemic uncertainty can be quite easily blurred.

Another example, albeit a little rote, is the classic example of rolling a die.[2] This is the prime example of aleatoric uncertainty. When the die leaves your hand, it is impossible to predict where it will land. You can calculate that the actual probability of rolling on any one side is 1/6, but it is impossible to predict where it will land as soon as it leaves your hand.

However, if you had perfect information about how the die was rolled, and a perfect physical model of rolling dice, you might be able to predict where it will land. Hence, someone could argue that this is epistemic uncertainty. However, someone else might argue that once a die is in the air, then the uncertainty is epistemic (given that we could build a physical model of the world to see where it will land), but while it is in someone's

---

[2] In the probability theory, examples using games of chance are very common. There are two reasons for this. First, as you will see later, the origins of the probability theory lie in gambling. Second, the probability theory presents an idealized environment to understand uncertainty, as people make up the rules of the game, instead of the real world, where the rules are made up by nature.

hand, it is aleatoric, since we wouldn't be able to have access to someone's nervous system and predict exactly how their hand will roll the die. As you can see, this argument becomes very complicated very quickly, and it is indicative of the kind of discussions that have shaped the study of uncertainty.

While someone might contend that all sources of uncertainty are epistemic, because it is within the capabilities of the human mind to know everything, we know that in practice this is definitely *not* the case. Plus, there might be some unforeseen limits to our knowledge, imposed by the nature of the universe—such asl Heisenberg's uncertainty principle (which I return to later).

Aleatoric and epistemic uncertainty can also coexist in a given problem. For example, imagine that you are playing roulette in a casino. The casino has two roulette wheels (that's probably a very small casino, but it's good enough for the purposes here). You have been informed that one of the wheels is malfunctioning and the distribution of outcomes is not uniform. Whereas in a standard wheel of 36 numbers, each number has 1/36 probability (roughly 2.7 percent), in the malfunctioning wheel, the ball has a 5 percent probability of falling on 0.

You are facing both kinds of uncertainty at once. First of all, you are facing epistemic uncertainty around which wheel is malfunctioning. But let's assume that the casino shares this knowledge with you, and now you know which wheel is malfunctioning. You still face aleatoric uncertainty, as the roulette follows the rules of probability. You cannot collect more data that will help you determine with 100 percent accuracy where the ball will fall next.

The exact type of uncertainty you're facing might change over time, depending on technological advancements. But this distinction between aleatoric and epistemic uncertainty has served us well so far.

And that's not where the confusion ends. Norvig and Russell's 1994 book *AI: A Modern Approach* also mentions *existence uncertainty* and *identity uncertainty*.[3] These are two kinds of uncertainty encountered in AI. Norvig and Russell give the example of an online book store that is using ISBNs (International Standard Book Numbers) to name each book. However, a book might have multiple ISBNs depending on editions and publishers over the years. In this case, the retailer faces existence uncertainty, as to what book each ISBN is actually referring to.

---

[3] `aima.cs.berkeley.edu/`, page 541

Identity uncertainty revolves around the identity of the customers. If anyone can register as a customer and leave a review of a hotel or a restaurant, then how can a new client be 100 percent sure that each registered customer is a different entity, and that a competitor chain didn't create 100 accounts just to leave negative reviews?

While these distinctions might be useful for researchers in the field of AI, for our purposes, these kinds of uncertainty can be neatly grouped under epistemic uncertainty. If we had more knowledge about the world, then both existence and identity uncertainty would simply disappear.

Now let's move on and explore one of the most fundamental problems of uncertainty, and how philosophers—from ancient history to modern times—approached the topic.

CHAPTER 2

Truth, Logic, and the Problem of Induction

"If it was so, it might be; and if it were so, it would be; but as it isn't, it ain't. That's logic."

—Lewis Carroll, *Through the Looking-Glass*

Imagine life in an ancient world. You are a hunter-gatherer. Sometimes your food consists of an animal you managed to hunt and kill, but other times it might be foraged fruits, nuts, and vegetables. Every day might bring forth a new plant, or a new mineral for you to eat. But how do you know what will provide sustenance, and what might poison and kill you and your tribe?

Well, the most basic thing you could do would be to simply find out which foods work best through trial-and-error. You simply eat whatever looks fine to you, and, if you survive, you can assume that the food is probably safe to eat. But can you be certain that this food will *always* be safe to eat, under all circumstances? What about foods that look similar to this one?

You are facing a situation of extreme uncertainty, which you need to handle—your life literally depends on it. Although you might not be conscious of it, you will go through a two-step thought process. The way you handle this situation is through the following process:

First, experimentation: You try it, see what happens, and hopefully don't die.

Second—all being well—you come up with some conclusions (and therefore theories) based on what you experienced.

After you try a particular kind of red berry ten times and you survive, you conclude that this berry is safe. You might then come up with a theory that *all* berries are safe. Or that all red fruits are safe to eat.

S. Kampakis, *Predicting the Unknown*, https://doi.org/10.1007/978-1-4842-9505-2_2

This is *inductive reasoning*. Induction is the process of collecting data, drawing conclusions, and coming up with hypotheses and theories that generalize from that data. Our ability to do this is one of the things that characterizes humans. In its simplest form (and the problem of induction can get very, very complicated—blame David Hume), inductive reasoning is the process of reasoning from the *specific* to the *general*.[1] An example of inductive reasoning is as follows:

> **I visited a new city and all the people I've seen so far wear black.**
>
> **Therefore, all the people in the city must wear black.**

The opposite of inductive reasoning is *deductive reasoning*—broadly speaking, you move from the *general* to the *specific*. The following is a well-worn example of induction:

> **All men are mortal.**
>
> **Socrates is a man.**
>
> **Therefore, Socrates is mortal.**

As these examples hopefully show, inductive reasoning is not always exactly watertight; it has baffled philosophers and mathematicians for centuries. How can you go from a set of data to a general theory and know that you are correct?

You're trying to come up with a way to fight uncertainty in the world, or at least a theory that might help you navigate uncertainty, but your reasoning faces uncertainty of its own.

Essentially, the difference between deduction and induction is that, given the correct premises, deduction will always give correct conclusions. But even given correct premises, induction might still give an incorrect conclusion.

Aristotle was the first person to systematically study deduction and induction. He described induction as an argument from the "particular" to the "universal." His focus though was on deductive reasoning and syllogisms, which form the basis for Aristotelian logic.

---

[1] This definition is considered outdated by many philosophers, but philosophical debate aside, it is good enough for the average layperson and for the purposes of this book. The curious reader will find lots of resources online about this topic, such as at places like the Internet Encyclopedia of Philosophy: `www.iep.utm.edu/ded-ind/`.

It was Francis Bacon who first studied induction in depth in his book *Novum Organum*, published in 1620. Bacon's method consisted of collecting data from the natural world through observation and experimentation and iteratively refining hypotheses.

John Locke also played a prominent role in the history of induction, as well as pioneering empiricism. According to empiricism, all knowledge is received through experience and the senses. In *An Essay Concerning Human Understanding*, Locke came up with the term "tabula rasa," which is the theory that all humans come into the world like a blank slate, without any prior knowledge. In other words, there are no innate ideas.

In the same work, Locke deals with the problem of induction. He writes:[2]

"Deny not but a man accustomed to rational and regular experiments shall be able to see further into the nature of bodies and guess righter at their yet unknown properties than one who is a stranger to them; but ... this is but judgment and opinion and not knowledge and certainty [...] the degree of assent we give to any proposition should depend upon the ground of probability in its favor."

This is an important statement. First, Locke recognized the issue of uncertainty in induction. While we can go from special cases to a general one, how can we be sure that our theory is right? Then, he uses the word "probability." While the theory of probability was not fully developed yet, Locke's use of the word here clearly demonstrates his opinion that any kind of induction is uncertain.

In spite of its usefulness, induction can look like a logical leap. We go from examples to a general rule, but how do we know that this rule holds? What happens in cases where we find a counter-example that disproves our theory? This is what Locke observed. Induction doesn't seem to follow from necessity. We can never be completely certain about our conclusions.

Inductive reasoning is directly connected to how we extract scientific truths, so its fallibility has widespread consequences. Just think about how much we rely on science to predict everyday occurrences—from a plane staying airborne to the sun rising.

So, what can be done?

Before I address this question, which is very much at the heart of *uncertainty*, let's take a look at some black swans.

---

[2] Shoutir Kishore Chatterjee (2003), *Statistical Thought: A Perspective and History*

The First Black Swan

"The problem with experts is that they do not know what they do not know."

—Nassim Nicholas Taleb, *The Black Swan: The Impact of the Highly Improbable*

Let's say that there's a new exotic creature that has appeared in the realm of zoology, called a "swan." This is pretty interesting—no one has ever heard of *swans* before. The few people who have encountered swans say that all of them have been white.

You are a swan enthusiast and, charmed by the stories you've heard about these magnificent creatures, you decide to study them more carefully. You embark on a journey to take as many photos of swans as possible. After years of laborious efforts, you capture 1,000 pictures of swans, all of them white. Can you conclude that all swans on the planet are white? How certain are you? Can you express this certainty in a probability, p_1, from 0 to 1?

How much would your certainty level change if you had taken 10,000 photographs, or a million photographs? Can you come up with a new probability, p_2? Would p_1 be smaller or larger than p_2? If you're like most people, you would—reasonably—assume that ten thousand or a million photographs give you more evidence and greater confidence than only 1,000 photographs. So, $p_2 > p_1$.

Let's say that all swanologists in the world for ten years have recorded only pictures of white swans. So, the consensus is that swans are white.

But then someone shows up with a picture of a black swan.[3] What now? *Now* the probability of the statement "all swans are white" is 0. Pure zero. Absolute certainty about the negation of a fact.

Regardless of how much you believed the white swans statement to be correct, you've been proven totally wrong. At least now, you have complete certainty about the opposite of this statement—"NOT all swans are white."

What this problem demonstrates is that all it takes is one counter-example for an argument that's been built by inductive reasoning to crumble to dust—no matter if we had billions of data points before that supported our argument.

---

[3] For the sake of argument, let's assume that computers and Photoshop have not been devised yet.

A similar form of this problem was first posed by the Scottish philosopher David Hume. Hume was the philosopher to systematically study the "problem of induction." While Locke was the first one to outline some of the issues of inductive reasoning, it was Hume who focused in-depth on the problem itself.

Also an empiricist, Hume posed the following situation. Let's say that I have observed that all particular instances of a bread (that look similar) have been nourishing. How can I *know* that the next instance will be nourishing and won't poison me?

Hume realized that just because we've observed a relationship of cause and effect in every instance until now, we cannot logically conclude that this relationship will prevail, and will always be true. This problem has long since plagued philosophy. The renowned 20th century philosopher Bertrand Russell wrote that if Hume's problem can't be solved, "there is no intellectual difference between sanity and insanity."[4]

Russell had a valid point. We can't fight uncertainty unless we have an infallible procedure that we can extrapolate, from our current data and logical premises, to more general theories of the world. However, regardless of the logical challenge that this problem presents, it is clear that humans have used inductive reasoning for a very long time, very successfully. Thus there are hints that—at least in practice, if not in theory—this problem can be solved. Let's take a look at how a few key figures addressed it.

The economist John Maynard Keynes, in his work, *A Treatise on Probability* (1921), proposed the following solution:

> "[...] the objects in the field, over which our generalizations
> extend, do not have an infinite number of independent qualities;
> that, in other words, their characteristics, however numerous,
> cohere together in groups of invariable connection, which are
> finite in number."

This has been called the "principle of limited independent variety." Keynes argues that the more instances we observe that verify a given hypothesis, the more certain we can be that the hypothesis can be true. This is a probabilistic approach to the problem. According to Keynes, after seeing 10,000 swans, we might conclude that the probability of the conclusion that "all swans are white" is close to 99.9 percent. There is still a chance that we might observe a non-white swan.

---

[4] Sir Bertrand Russell, *A History of Western Philosophy* (1946)

Bertrand Russell takes a more pragmatic approach. According to his view, the world is characterized by certain kinds of regularities. Living things form habits based on those regularities, by performing inference, which in turn helps their adaptation and survival in the natural world. Russell asserts that:

> "The forming of inferential habits which lead to true expectations is part of the adaptation to the environment upon which biological survival depends."

Therefore, according to Russell, the fact that evolution has empirically validated this approach is enough justification to explain why inductive reasoning is valid. It's an interesting view, partly because it is so closely related to machine learning, as you'll see in a later chapter.

Karl Popper is one of the most famous names in epistemology. Popper concedes that inductive reasoning is false—we can never be completely certain that our theories are correct. However, we *can* be certain that our theories are false. The only thing we need to falsify a theory is a counter-example; in this instance, the exception *dis*proves the rule.

According to Popper, what we should do is accept theories provisionally until they are falsified, and then come up with new ones. Progress in science is achieved by the constant falsification of theories and the creation of new ones. Popper calls this the "deductive method of testing."

This approach might seem pessimistic, but it is sound from a logical perspective. From an empirical perspective, however, not many people think this way, and in practice the theory has received criticism.

The great contribution of Popper to science is that he gave us the minimum requirement for a theory to be considered scientific. According to his "principle of falsifiability"—also known as "Popper's falsification principle"—scientific theories can only be considered scientific as long as they are falsifiable.

Let's say for example that you pose the following theory:

> *While the sun rises and sets according to the laws of planetary motion, the rise and fall of the sun is also influenced by a unicorn which lives in Alaska. The unicorn, called Bob, needs to use his shiny, spiraling horn, to make the sun rise. Bob is invisible, and he doesn't like humans too much. Hence, you are unlikely to be able to find him, even given the fact that he lives in Alaska.*

This is similar to the theories that many ancient people came by as to why the sun rises. Some supernatural entity has control over heavenly bodies and physical phenomena, like rain and tornadoes.

So, what's the issue with Bob the unicorn, or with deities that control the weather? The issue is that there is no way to prove whether these theories are correct or not. What if the unicorn was named Sandra, instead of Bob? Would this make a difference? How can you test any hypothesis around this theory? You simply can't.

There have been similar theories that have been devised, which under close inspection, are not falsifiable, and must be rendered pseudo-scientific.

The most popular theory that has been attacked on those grounds is Freud's psychoanalysis. While psychoanalytic practice is still popular in many places around the world, a large part of the psychoanalytic interpretations of the mind are considered non-scientific. This is due to circular arguments, and no clear path to creating an experiment that can falsify those theories.

For example, let's say that you have been diagnosed with some form of neuroticism by a psychoanalyst. You object to this, and then the psychoanalyst responds, that the reason you object is because you are neurotic. This is an example of circular reasoning. If anyone who disagrees with the characterization of neuroticism does so because of their neuroticism, then how can we disprove the theory? Well, we can't.

Another example of an issue with psychoanalysis is Freud's theory of interpretation of dreams. According to Freud, the dreams are symbols for repressed urges, which, if left unleashed in the true form into our dreams, they would terrify the subject. These urges can be very animalistic, violent, and sexual in nature. A psychoanalyst has to find interpretation of the various symbols used in dreams. However, in studies that were done in the past, different psychoanalysts came up with radically different interpretations of those dreams.[5] If different experts can come up with radically different interpretations of those same dreams, how can we actually falsify this theory?

Karl Popper personally attacked psychoanalysis in his book *Conjectures and Refutations: The Growth of Scientific Knowledge* (1962). Popper was highly critical of both Freud and Alfred Adler. Adler was one of Freud's first disciples, and Popper had the chance to work briefly in one of his clinics. Popper argued that the main issue with psychoanalysis was that it could accommodate and explain all of human behavior, but it couldn't really offer any means to falsify it. There are no real predictions, only ad hoc explanations.

---

[5] Matthew Walker in his book *Why We Sleep* offers a full account of the science of dreams.

In spite of the criticism, psychoanalysis is clearly still around. While Popper's criticism is true to some extent, it also misses some other important points. Freud's psychoanalytic theory also brought to the foreground some important aspects of human psychology, such as the existence of the subconscious. Secondly, Freud was one of the first people to study the development of children. He found explanations for why children act in certain ways, which to the eyes of an adult, may seem baffling.

The story behind Popper's criticism of psychoanalysis brings to the forefront an important issue around uncertainty. Psychoanalysis attempts to explain many things, like dreams, once they've occurred. But how can we know that those explanations are correct? This isn't just a question of psychoanalysis; we encounter it in our daily lives as well. We all have theories as to why people make certain decisions, or how our decisions can affect others. New parents face these dilemmas all the time and come up with ever-evolving theories as to how to raise their children.

But it's not easy to be certain about the exact cause and effect of an event, *after* something has happened. This is a common fallacy in—for example—studies of the best business practices. Every now and then, you might encounter an article about the habits or strategies of the most successful businesses and entrepreneurs. What these studies *don't* tell you is how many businesses and entrepreneurs followed the exact same principles but failed completely. By focusing only on positive examples after something happened, we might come up with all sorts of explanations that seem obvious. However, in reality, it is impossible to confirm any of our theories without having counter-examples. We need to compare different entrepreneurs and strategies, some of which failed and some of which succeeded, to find the deciding factor. Without variation, there can be no learning.

Thomas Kuhn, arguably one of the most influential epistemologists of all time, alongside Karl Popper, believed that progress is achieved through paradigm shifts. Contrary to the dominant thinking at the time, in his 1962 book *The Structure of Scientific Revolutions* Kuhn argued against the idea that progress was only achieved through the accumulation of more facts and knowledge.

According to Kuhn's understanding, a paradigm is a way through which we see the world and interpret knowledge. Scientific theories are not constantly improved through the addition of more facts. Rather, progress takes place through the replacement of old paradigms with new ones—a paradigm shift. This often takes place through the death of an older generation of scientists and the emergence of a new one.

For example, Kuhn analyzed the Copernican revolution. Nicolaus Copernicus was the mathematician and astronomer who placed the sun, rather than the earth, at the center of the universe. Before Copernicus, the Ptolemaic system, devised by Ptolemy, was accepted as the true version of the world, with earth at the center of the universe. Of course, now we know that we live in a much larger universe than our solar system, but it took more than 1,300 years for humanity to figure that out.

The Copernican revolution was not just an addition of facts, but a radically different way to view the world. But it was not until Galileo Galilei that the paradigm shift was complete. Copernicus was still using Ptolemy's tools of cycles and epicenters, which resulted in an overcomplicated model that no one believed in.

Galileo made the shift possible, and Johannes Kepler would later complete the paradigm shift by introducing the heliocentric system. The minutiae of these theories are not wholly relevant to our discussion; the main point here is that world-changing discoveries were made possible by rejecting past knowledge—out with the old, and in with the new. And "new" does not sit easily alongside "certainty."

Coming back to psychology, psychoanalysis gave us what is also called the "hydraulic" model of the mind. According to this model, subconscious urges built up, and then need to be released. The information age later gave us the computational model of the mind, according to which our mind is like a computer with a memory and a processing unit.

More recent analogies have been inspired by the successes in data science. Named after Thomas Bayes, the Bayesian approach believes that the brain is conducting some form of integration. This inspired an approach called "rational analysis." According to this view, the mind has evolved to be adapted to its environment. The inventor of the theory, John Anderson, suggests that we have to follow a six-step process to study it.[6] The traditional cognitive approach to the study of the mind assumes that humans have biases; for instance, one might assume that all Liverpudlians (natives of Liverpool) are funny, based on past experiences meeting Liverpudlians. Rational analysis assumes that humans are actually rational, given that the mind has been optimized to be adapted to its environment.

---

[6] 1. Goals: Precisely specify the goals of the cognitive system.
  2. Environment: Develop a formal model of the environment to which the system is adapted.
  3. Computational Limitations: Make the minimal assumptions about computational limitations.
  4. Optimization: Derive the optimal behavioral function given Steps 1-3.
  5. Data: Examine the empirical literature to see whether the predictions of the behavioral function are confirmed.
  6. Iteration: Repeat, iteratively refining the theory.

The move from the hydraulic model, to the computational model, to rational analysis represents a paradigm shift. One view assumes that the mind is composed of urges that require release; another model assumes that the mind is like a flawed computer. The latter approach suggests that the mind is actually optimized for its environment, so we need to see it from a different lens.

According to Kuhn, different theories are "inconsummerable." This means that we can't understand one theory from the perspective of another theory because they have no common basis, or shared standard for comparison. You can't use tools from one paradigm to prove or disprove another one. The interesting corollary of this is that scientific progress is directionless.

The theoretical physicist Steven Weinberg captured this idea:[7]

> "Kuhn recognizes that Maxwell's and Einstein's theories are better
> than those that preceded them, in the same way that mammals
> turned out to be better than dinosaurs at surviving the effects
> of comet impacts, but when new problems arise they will be
> replaced by new theories that are better at solving those problems,
> and so on, with no overall improvement."

Regarding the problem of induction, Kuhn refutes Popper's approach. He says that "There is no such thing as research without counterinstances [...] If any and every failure to fit were ground for theory rejection, all theories ought to be rejected at all times." According to Kuhn, science is not a truth-seeking enterprise. Instead, theories are in some sort of evolutionary game, where they compete with each other and the best ones survive.

We could characterize Kuhn's approach as pragmatic and radically different from the approach of his predecessors. He recognizes the human component in the scientific process. However, his theory has been criticized, as not all progress in science takes place through revolutions.

Another kind of uncertainty emerges in this case. This is the uncertainty around scientific theories. Science has gone through various degrees of popularity across the ages. From the solution to all of humanity's problems, we currently live in an era where experts are discredited and fake news floods the web. Just think of the number of people who don't believe in global warming.

---

[7] Steven Weinberg (2001), *Facing Up*

The history of inductive reasoning highlights that, when fighting against the unknown, there are no easy solutions. Those who consider science a panacea often forget that it is a *process* with many of its own difficulties, bumps, and hurdles.

Even if we can never logically conclude that inductive reasoning is correct, maybe the challenge is coming up with the answer as to why it seems to be right in so many cases.

CHAPTER 3

Swans and Space Invaders

"Mediocristan is where we must endure the tyranny of the collective, the routine, the obvious, and the predicted; Extremistan is where we are subjected to the tyranny of the singular, the accidental, the unseen, and the unpredicted."

—Nassim Nicholas Taleb,
The Black Swan: The Impact of the Highly Improbable

One of the most famous current scholars of uncertainty is Nassim Nicholas Taleb. Now a scholar, essayist, and author of the *Incerto* collection of works (of which *The Black Swan* is the second title), Taleb used to work in finance—arguably the industry most exposed to uncertainty. Fortunes are made and lost in the blink of an eye.

Taleb's view is that one of the main issues around finance and economics is that they belong to what he calls "Mediocristan" (think of it as a place); on the other hand, much of our world belongs to "Extremistan," What Taleb refers to as Mediocristan has dominated Western thought to a large extent. Mediocristan describes statistical thought that is concentrated around averages and the "normal" distribution. As learned saw earlier, that method of reasoning has important consequences.

Much of our world is characterized by normality. You can see what a normal distribution looks like in Figure 3-1.

© Stylianos Kampakis 2023
S. Kampakis, *Predicting the Unknown*, https://doi.org/10.1007/978-1-4842-9505-2_3

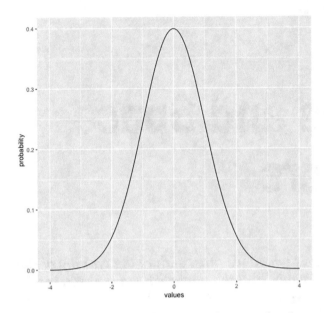

Figure 3-1. *The standard normal distribution. The standard normal distribution has a mean of 0 and a standard deviation of 1. It is used in countless mathematical models*

The normal distribution has an important property: 95 percent of the values are within one standard deviation of the mean, and 99 percent of the values are within two standard deviations from the mean. A good example of a pattern that follows the normal distribution curve is height—within a certain age group and a given culture, we don't expect to see many people who are very short or very tall.

IQ also follows the normal distribution, as well as weight, and many other physical phenomena. The probability of observing someone who is four meters (13 feet) tall or who weighs 1 ton (2,200 pounds) is practically zero. This also corresponds closely to our everyday experience.

The normal distribution is used by the most popular statistical model of all time: linear regression. Other probability distributions have been used for similar models, with all the distributions similarly discounting extreme values. Many statistical models of this kind have successfully been employed in fields as diverse as education and ambulance waiting times.

This is what Taleb refers to as "Mediocristan." With models in Mediocristan, extreme values are not only unlikely, they're also not going to influence the model too much. For example, in statistics it's a common practice to remove outliers, just to make the results a bit clearer. Outliers are data points that do not conform to the average case. An example is shown in Figure 3-2. This plot shows average total sales against the day of the week, for a fictional enterprise. The fifth point is obviously a clear outlier, as it is so much higher than the average sale.

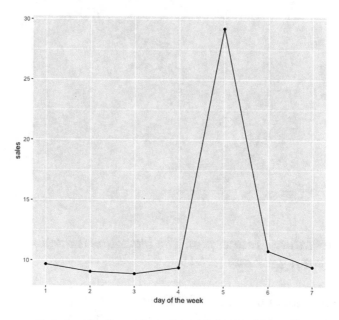

Figure 3-2. *Average total sales based on day of the week, with a clear outlier*

Let's say that you are conducting a study on the effects of a city's temperature on the consumption of ice cream cones. The two follow a close linear relationship—the higher the temperature, the more ice cream cones are consumed. But let's imagine that there was a one-off ice cream festival during the month of December, and the best ice cream producers in the world flooded into the city in a huge celebration of ice cream. Consumption of ice cream reached a high during the month of December, even though temperatures were at their lowest. This is an outlier. It doesn't describe the general—and normal—relationship between temperature and ice cream consumption. Hence, you can remove it from the dataset.

But an interesting property in Mediocristan is that even when we include the outliers, they won't radically change the results. Let's see in this imaginary example what a plot looks like without outliers. You can see this in Figure 3-3.

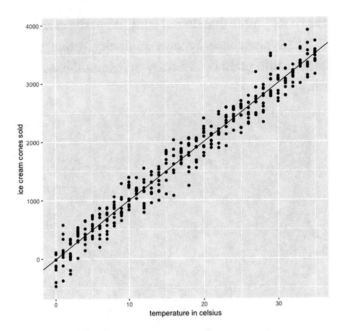

Figure 3-3. *Temperature vs cones plot. The black line is the line of best fit, produced by a linear regression model*

The slope of the regression line is 100. The slope in a linear regression model measures the strength and direction of the relationship. In this model, this means that for 1 degree Celsius increase in temperature, consumption of ice cream cones increased by 100.

Let's see what the model looks like *with* outliers. We have 15 days where the temperature was 0, but the consumption of ice cream was very high. This is shown in Figure 3-4.

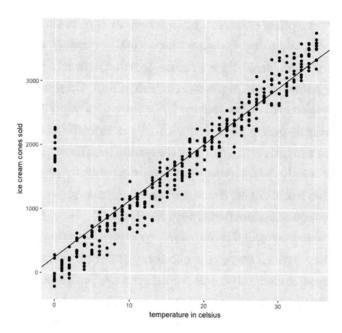

Figure 3-4. *Temperature vs cones plot and outliers*

The 15 days are clearly visible. Consumption was abnormally high for those days. But the overall trend is still linear. The actual slope of the model is 90. This doesn't differ radically from the previous model.

Taleb gives the following example of Mediocristan in *Black Swan*:

> "Assume you round up a thousand people randomly selected
> from the general public and have them stand next to each other
> in a stadium. [...] Imagine the heaviest person you can think of
> and add him to that sample. Assuming he weighs three times the
> average, between four hundred and five hundred pounds, he will
> rarely represent more than a very small fraction of the weight of
> the entire population (in this case, about a half a percent.) [...]
> You can get even more aggressive. If you picked the heaviest
> biologically possible human on the planet [...] he would not
> represent more than, say, 0.6 percent of the total, a very negligible
> increase."

Hence, extreme values are uncommon, and even when they show up, they won't drastically alter our models and our opinion about the world.

Extremistan, on the other hand, is a place where extreme values are prevalent. A classic example of this is income. Imagine that we take a random sample of people from the United States and write down their total net worth. On one hand we have people like Bill Gates who, at the time of writing, is worth $100 billion. On the other hand, we have a number of people who are on the average U.S. salary of around $40-$50k. This time, a single person controls 99 percent of the wealth. The distribution is completely unequal. However, this doesn't surprise us. We know that some people are extremely poor, and some others are extremely rich. This kind of dynamic exists in many systems. But we often confuse the two worlds of Mediocristan and Extermistan—and when we're caught by surprise, the consequences are not always favorable.

According to Taleb, traditional theories in economics and finance are ridden with models from Mediocristan. However, the economy and many other arenas of global scale are from the land of Extremistan. Extreme events will take place regularly, and they can have profound effects on the system. According to Taleb, a black swan is an outlier, but an outlier with a massive impact. While nothing in the past could point to it being likely to happen, we come up with explanations *after* the fact that make it seem predictable.

Wars and financial crises are some of the best examples of black swans. Taleb was critical of the financial industry and its practices even before the great financial meltdown of 2008, and he essentially predicted the crash (something that would shoot him to fame). After every financial crisis, multiple theories are offered as to why and how they happened, but financial crises have not stopped taking place. Similarly, anyone can find multiple explanations as to how World War I started. The official start of the war is often considered to be in 1914 when Gavrilio Princip assassinated Franz Ferdinand. Historians also provide various other geopolitical and historical reasons that led to the start of WWI.

These explanations might make the war look like a natural outcome of the events that preceded it. However, they do not take into account two things. First, the actual assassination was a random event that could have gone either way. A detailed account of the event shows that the assassination was far from easy. It could have been that Franz Ferdinand survived. Secondly, all these explanations come *after* the event has taken place.

According to Taleb, the main issue with black swans is that they are unpredictable and have a huge impact, but many people in important positions foolishly believe that we are either immune to them or that we can predict them. A good example to see how this works in real life is the example of the turkey before Thanksgiving, given by Taleb.

> "A turkey is fed for a thousand days by a butcher; every day confirms to its staff of analysts that butchers love turkeys "with increased statistical confidence." The butcher will keep feeding the turkey until a few days before Thanksgiving... [The] turkey will have a revision of belief – right when its confidence in the statement that the butcher loves turkeys is maximal and 'it is very quiet' and soothingly predictable in the life of the turkey."

So, many people in the real world are a bit like this turkey. They make predictions based on past experience, unaware of other events unfolding in the world around them, with potential catastrophic consequences. Taleb notes that 99 percent of the total variation in the value of derivative portfolios between 1988 and 2008 occurred in one day, when the European Monetary System collapsed in 1992. A single event can have a larger impact than a number of small ones. Weight belongs in Mediocristan. It's impossible to lose lots of weight in a single day. But money belongs in Extremistan. It is entirely possible to lose lots of money within a single hour, even in a minute.

As the world grows more interconnected and technology progresses, black swan events enter our lives more frequently. A single war these days could cost more lives than all the wars during the middle ages combined. A large company defaulting in one country can cause a global meltdown.

Taleb takes a very comprehensive view of uncertainty and provides a good description of how it can enter our lives. From personal events, to social and economic events, we are all exposed to the possibility of an unpredictable dramatic event entering our lives. Taleb is critical of various tools employed to study uncertainty, as many of them come from Mediocristan. He is somewhat pessimistic about our ability to predict black swans, or the usefulness of even trying to do so.

His suggestion is to take a more strategic approach. His trading strategy consisted of a large number of very safe bets, and a small number of bets that could potentially be very profitable. Depending on your goals, he suggests a similar strategy. The first thing to do is recognize that you live in a world where black swans can take place. Secondly, in your work, you want to be in a position where you can benefit from positive black swans. If you are an entrepreneur, you might go to a conference, where you meet an investor that can skyrocket your idea.

Taleb makes two major contributions to the study of uncertainty.

- First, he provides a very good critique of the way that many statistical methods are misused in industries, where, by their very nature, black swans can happen and can have disastrous consequences.

- Secondly, he proposes a qualitative strategy, which can be used in real life. He suggests to try to balance risk following the "barbell" strategy, named in this way, because a barbell has two modes (two extremes). Let's say that you are thinking of investing your savings. Taleb suggests that you invest 90 percent of your savings in some very low risk, boring investment, with very small probability of failure. Ten percent of your investments should go to high risk investments, where black swan events might actually benefit you.

However, Taleb might be more pessimistic than he should be. One argument against Taleb is that through more and more data collection, humans might actually be able to predict some of the events that are currently unpredictable. Chapter 1 discussed the different kinds of uncertainty: aleatoric and epistemic. Many black swans in our world might be down to epistemic uncertainty. An example of a black swan event is a natural disaster. However, more accurate scientific methods and a greater understanding of global warming will help humans predict them with greater accuracy before such events, like tsunamis, take place.

Secondly, we do have some control over social systems and how they are structured. The financial system has a long and complex history, but it was made by humans, who made a sequence of decisions across the centuries in order to arrive at its current form. Obviously, efforts to completely control society are neither condoned nor have they proven successful (consider the Soviet Union, and many dictatorships across the world). However, the rules of the game that generate black swans are changing. This major shift of globalization seems to have caused massive inequalities and large populations of winners and losers.

Occam's Razor, Space Invaders, and Lizard People

"Reports that say that something hasn't happened are always interesting to me, because as we know, there are known knowns; there are things we know we know. We also know there are known unknowns; that is to say, we know there are some things we do not know. But there are also unknown unknowns—the ones we don't know we don't know. [...] The absence of evidence is not evidence of absence, or vice versa."

—Donald Rumsfeld

There may never be a better quote than this to capture the implications of decision-making under uncertainty. Donald Rumsfeld was the secretary of defense under George Bush Jr. from 2001 until 2006. Rumsfeld gave this answer when he was asked whether the United States had information about Saddam Hussein selling weapons of mass destruction to terrorist groups. We constantly face decisions that are shrouded in uncertainty, with bigger and smaller consequences. For Rumsfeld and Bush's government, their decision was, of course, catastrophically massive.

When faced with uncertainty or a shortage of data, we might come up with all sorts of different hypotheses. But how can we choose among different ones? Is it more likely that someone is hiding weapons of mass destruction or not? Is your spouse cheating on you? Is there some secret conspiracy controlling a large part of humanity? We need a rule to help us figure this out.

One of the most popular heuristics in this regard is Occam's razor. Occam's razor is often wrongly attributed to William of Ockham (1287–1347), who was a philosopher and theologian from Surrey, England. The concept had been around for centuries before Ockham, and the term Occam's razor was actually coined later on. However, Ockham was famous for using it, so it was associated with him (as well of course as the obvious similarity in nomenclature!). It was John Punch (1603–1661) who came up with the most famous definition of Occam's razor:

> *Non sunt multiplicanda entia sine necessitate.*

> *Entities are not to be multiplied without necessity.*

In other (and simpler) words, Occam's razor can be formulated as follows:

> "Among all things being equal, the simplest explanation tends to
> be the best."[1]

This is a heuristic that has served us well for a long time. From statisticians to detectives, when faced with multiple conflicting hypotheses, the simplest one is often true. But quite often we forget this simple rule.

For example, your spouse might have not returned your messages for the last couple of hours. Maybe they are busy at work, or have they flown off to some remote island with their lover? Or maybe something tragic happened to them and they are in hospital?

You got a flat tire driving back home. Is it more likely that the tire was defective, or that a sniper shot your tire so that you would be late home and they would have more time to steal your TV?

The outcome of an election is often not what most people expected. Would you say that sometimes forecasters and pollsters get predictions wrong, or that a conspiracy of subterranean lizard people has changed the outcome of the election?

As you can see, in all these examples, both explanations are possible (with, perhaps, the exception of the last). However, only one of the two explanations is *plausible*. A popular saying that relates to Occam's razor is, "extraordinary claims demand extraordinary proof."

Apt examples are found in studies in parapsychology. There are multiple studies done on parapsychological phenomena, from telepathy to precognition. At the time of writing, the term "parapsychology" returns 33,000 results on Google Scholar. Some of those studies even gave statistically significant results. So why isn't parapsychology mainstream? Theories like telepathy are disruptive of our worldview. There is no current model of physics to clearly support how something like that would work. Hence, the existence of telepathy constitutes an extraordinary claim, and we would need extraordinary proof in order to back it up.

---

[1] "Entia non sunt multiplicanda praeter necessitatem" in Latin.

The physicist Stephen Hawking advocated for Occam's razor in *A Brief History of Time*:

> "We could still imagine that there is a set of laws that determines events completely for some supernatural being, who could observe the present state of the universe without disturbing it. However, such models of the universe are not of much interest to us mortals. It seems better to employ the principle known as Occam's razor and cut out all the features of the theory that cannot be observed."

But why does Occam's razor work? David MacKay, a British physicist and machine learning researcher, gave an account of this in his book *Information Theory, Inference, and Learning Algorithms*. Occam's razor obeys the rules of Bayesian inference. What does this mean?

McKay gives the following example. Let's say that you have the following sequence of numbers:

$$-1, 3, 7, 11$$

Which of the following rules best predicts the next digit?

Rule 1: Add 4 to the previous number.

Rule 2: Let X be the previous number. Evaluate the next number by the formula $-\dfrac{x^3}{11} + \dfrac{9}{11x^2} + \dfrac{23}{11}$.

If you run the numbers, rule 2 also correctly predicts this sequence. However, it looks overly complicated.

So, how can Occam's razor help us choose a rule in this case? Let's say that someone gives you a hint and tells you the numbers in this problem can only be between -50 and 50. The probability of observing this data under Rule 1 is 0.0001. The probability of observing this data under Rule 2 is around 2.5×10^{-12}. So, the first rule assigns a higher probability to what we observe than the second rule. Hence, the first rule is a plausible choice.

The intuition behind what happens is that more complicated theories can explain everything in principle. When this is translated to probabilities, the fact that they explain everything means that each possible event is actually given a very low probability. On the other hand, simpler and more focused theories (Rule 1 in this case) assign larger probabilities to the observed evidence.

In short: A theory that tries to explain everything, usually ends up explaining nothing.

Occam's razor is also a great heuristic that explains why most conspiracy theories are wrong. Believing that the world is secretly controlled by a reptilian elite requires one to accept multiple complicated theories:[2] That the existence of reptilian humanoids is possible; that they are intelligent, and so intelligent in fact, that they can control humanity; and that they have been extremely successful in keeping their existence a secret. Or we can just believe that they don't exist—which is the simplest explanation. Likewise, we can accept that the earth is flat. But, then how can we explain airplanes travelling around the world? Maybe they simply travel to the edge of the world and teleport on the other side (like in a video game), or the pilots are part of the conspiracy too. Or maybe, the whole flat earth idea simply doesn't make sense.

Another reason that conspiracy theories are unlikely is that many of them are not falsifiable. If a theory is too complex and flexible, then it can easily accommodate any data or exceptions. However, this makes it a "bad theory." Coming back to the reptilian example, why have you not encountered these creatures in real life? In defense of the theory, someone might argue that these creatures have shape-shifting capabilities. Or that we are brainwashed with radio waves, so that they are invisible to us. All these are patches that simply add to the complexity of the original theory, making it completely unfeasible.

So, is Occam's razor a perfect heuristic? Well, no. It is a heuristic after all, and there are some cases in which it will fail. Coming back to Donald Rumsfeld, the unknown unknowns are basically what Taleb describes as black swans. These are events that we can't predict, but that could have massive consequences. While Occam's razor could lead us to take them out of our models, this doesn't mean that we can't prepare for them, given that the resources, or energy, we spend to prepare do not exceed a reasonable limit.

Let's consider an example. Many people believed that the end of the world would arrive in 2012. This was because 2012 was the end-date of the calendar of the Mayan civilization. Some individuals took drastic measures to protect themselves against

---

[2] u.osu.edu/vanzandt/2018/04/18/the-world-is-controlled-by-a-group-of-elite-reptiles/comment-page-1/

the impending end of the world. More than 100,000 people were planning a trip to a mountain in France which was supposed to be able to withstand the end of the world. The *Independent* reported on the incident[3]:

> "A grizzled man wearing a white linen smock, who calls himself Jean, set up a yurt in the forest a couple of years ago to prepare for the earth's demise. 'The apocalypse we believe in is the end of a certain world and the beginning of another,' he offers. 'A new spiritual world. The year 2012 is the end of a cycle of suffering. Bugarach is one of the major chakras of the earth, a place devoted to welcoming the energies of tomorrow.'"

Even NASA published an article explaining why the world *would* end in 2012.[4] Other people went so far as to build bunkers to protect themselves against the impending doom.[5]

So, what is wrong with all this from an Occam's razor perspective? First of all, some of the theories employed as to the reasons behind the end of days are overtly too complicated. Invoking chakras or other mystical energies as the reason of some impending global disaster is a more esoteric explanation than necessary. A simpler explanation is that an ancient civilization that lived thousands of years ago wouldn't have the means to predict anything of significance that could have happened in 2012.

Secondly, let's assume for a second that the world might indeed end. It is true that, at least in theory, a comet could hit the earth, ending life as we know it. Or a war could break out, leading to nuclear destruction, obliteration of major cities, and huge geographical areas being uninhabitable. Given that this is indeed a possibility, even a remote one, a bunker might not be an entirely unreasonable idea for someone who has the resources to build one. If you are of average salary, then worrying too much about an event that has a probability of taking place smaller than 0.00000000001 is probably not worth it. But given adequate resources, you might have the luxury of worrying about any kind of problem that has a non-zero probability of taking place.

---

[3] www.independent.co.uk/news/world/europe/hippies-head-for-noah-s-ark-queue-here-for-rescue-aboard-alien-spaceship-7584492.html

[4] www.nasa.gov/topics/earth/features/2012.html

[5] www.huffpost.com/entry/mayan-apocalypse-bunkers-popular_n_2294826

Indeed, when having to deal with black swans (the unknown unknowns), it is our priorities and resources that decide whether it is wise to spend any time contemplating them.

First of all, a simple solution doesn't mean it's correct simply due to its simplicity. Some current theories in physics might be more complicated than previous ones. Most people will argue that quantum physics is admittedly more complicated than the theory that gods have created the universe. And in another instance, the assertion that homosexuality is wrong because it isn't biologically "simpler" is, equally, hugely flawed.

Secondly, simplicity is in the eye of the beholder. If suddenly all of humanity found itself, through some miracle, to possess an average IQ of 250, then whatever seems complicated now, would seem very simple then. What might be simple to an expert, might be complicated to a novice. There is no easy way to define simplicity, and as a concept, it can be more than vague.

For example, there was a concept in experimental psychology, very popular for decades, called "radical behaviorism." According to this, the mind is some sort of illusion, and concepts like thinking are useless for understanding humans. Instead, all we need to do is to understand *behavior*. Clearly, this approach eventually failed and gave rise to other theories, such as the computational approach to cognition, which might be more complex, but also can better explain a wider range of phenomena. On the other hand, Einstein's relativity equation of $E = mc^2$ is rather simple, so it seems to follow this rule.

The take-away is that Occam's razor is a heuristic, and heuristics are not always correct. The principle of parsimony is important in statistics and machine learning—as you will see later—but when dealing with theories, things might be vaguer.

Therefore, the wisest way to apply learning is as follows:

1) If two theories explain exactly the same data, then go for simplest one.

2) If new data falsifies the theory, then develop a new theory.

Simple, right? (Pun intended.)

These approaches are consistent with both the Bayesian view of Occam's razor, and also some of the prominent epistemological theories we explored earlier, such as Kuhn's and Popper's. In practice, this might look easy, but as you saw, concepts like simplicity can be quite vague. Also, while logical reasoning can take us up to a point, in order to go further, we need quantifiable measures. This has always been the case for any physical quantity, from weight, to temperature, to height. Uncertainty is such a quantity, and probability theory is the best way we have discovered so far for predicting it—until now, that is.

CHAPTER 4

Probability: To Bayes or Not To Bayes?

"Life is a school of probability."

—William Bagehot

What is probability?

If you're not a data scientist, you likely have an intuitive understanding of probability from school lessons or based on common sense. For many events, it is impossible to always correctly predict the outcome in advance. But we know that some outcomes are more likely than others. Put simply, probability is a way to express and study which outcomes are more, or less, likely to happen. The etymology of the word "probable" has an interesting bearing on how we use it today; it comes from the 14th century French word *probable,* meaning "provable or demonstrable," originally derived from the Latin verb *probare*—"to try, to test."

If I asked you what's the probability of getting tails on a coin flip on a fair coin, you would probably say 50 percent. Intuitively it makes sense. However, you might not be able to explain why this should be true mathematically. Coin-flipping is a simple case for calculating probabilities. How would you define the probability of Manchester United winning the next Champions League? You might have your own personal beliefs about which football team is the best (and this belief could change every year), but the way you come up with this probability "feels" different from the coin-flipping case.

If you're a firm believer in the fact that Manchester United will win the Champions League this season, you might say you give it a 70 percent probability. But how do you know it isn't 71 percent? Or 61 percent? In the coin-flipping scenario, if someone told

© Stylianos Kampakis 2023
S. Kampakis, *Predicting the Unknown*, https://doi.org/10.1007/978-1-4842-9505-2_4

you the probability of landing on heads is 51 percent, then you'd say the coin isn't fair. In a casino, a 1 percent change in probability in a game can be the difference between the casino making a huge fortune or incurring a huge loss.

We're all familiar with probability—it informs even the minutiae of our daily lives, such as the likelihood of a bus turning up on time—it's our main tool to express and assess uncertainty. Just think how many times you say "probably" in a week. Despite this, it's actually pretty tricky to define probability and to understand exactly how (and why) it works.

Frequentist or Bayesian?

The most popular definition of probability, and arguably the most intuitive, is the frequentist one (also known as *frequentism*). According to frequentists, an event's probability is defined as the limit of the event's frequency in a large number of trials.

What does this mean? Let's go back to the example of flipping a fair coin. You said that the probability of rolling heads on a single roll is 50 percent. However, how do you know this to be true? What if you roll tails ten times in a row? Would this change the probability of rolling heads? Obviously not. Intuitively, this makes sense, but why?

I ran a coin-tossing experiment (simulated in the R programming language[1]); you can see the results in Figure 4-1. The proportion of heads very quickly converges to 50 percent.

---

[1] The R language (www.r-project.org/) is the most popular language for statistics. Python, however, seems to have overtaken it for machine learning.

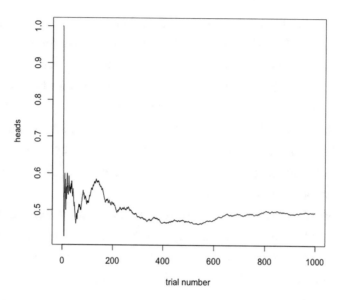

Figure 4-1. *Coin tossing experiment*

This is the definition of frequentist probability in practice. If you execute an experiment a large number of times, then the frequencies will converge to their true probabilities.

Frequentist statistics have been the orthodox branch of statistics for most of history. In his *Rhetoric*, Aristotle wrote that "the probable is that which for the most part happens." The practice of statistics is based on the belief that you can extract a sample from a population, and then study properties of the population. If we treat each entity in this sample as an experiment, then the more samples we collect, the closer we will get to the truth.

However, there are certain events for which the frequentist understanding makes no sense. Many of these events have seismic impacts on our lives—for example, the results of an election or referendum. We can't run an experiment of 1,000 elections and see what happens. And we can't run 1,000 Champions League finals to find the true probability of Manchester United winning it.

It's at this point where the Bayesian definition of probability comes to the rescue.

Though he never personally published his theories—they were written down and edited posthumously by his friend Richard Price—Reverend Thomas Bayes gives his name to Bayes' theorem, or Bayesian theory. Ironically, no one is certain exactly when Bayes was born (though he died in 1761), or what he looked like; one portrait survives, but it's uncertain that the man in it actually *is* Bayes.

Bayes used conditional probability (which you'll look at shortly) in *An Essay towards solving a Problem in the Doctrine of Chances*, which was presented to the Royal Society in 1763. Bayes' work concerned the following problem: How can you know the probability of an event, based only on how many times it occurred—or didn't occur—in the past? Bayes used a thought experiment to illustrate his argument.

Bayes has his back turned toward a table, and his assistant throws a ball on it. There is equal probability of the ball falling anywhere on the table. Bayes has to guess where the ball is. On the first throw, Bayes experiences the maximum degree of uncertainty, as the ball could really be anywhere on the table.

In the next step, his assistant throws another ball and reports whether it fell on the left or the right side of the first one. Let's say that this time the ball lands on the right side of the table. We can assume that the first ball now is more likely to be on the left side. If the original ball had landed on the left side, then the right side would have more space for another ball to land.

Then, the assistant throws another ball and this ball lands to the right again. This makes it even more likely that the original ball lies on the left. Hence, with each throw, we narrow down the position of the original ball more and more. This idea is illustrated in Figure 4-2.

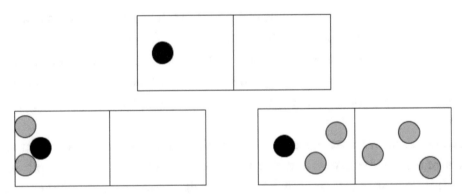

Figure 4-2. *Depiction of Bayes' argument.*

As shown in Figure 4-2, after the first ball is thrown (black ball), the second (orange ball) has more space on the right of the table than on the left. On the second row (left), you can see that there are only a few positions where the second ball could end up left of the original ball. On the second row (right) graph, you see all the positions right of the original ball, where the orange one could lie. It is clear that there is more space right of the original ball, and most of this space lies to the right of the table. However, it is still possible for the orange ball to be right of the black ball, but on the left side of the table. So, two ball throws are not enough to completely pinpoint the location of the black ball.

In the modern world, this might remind us a bit of the Battleships board game. In battleships, each player (the game has only two players) has to place ships on a square board. On each round, the players specify one point in the board that they attack. If they hit part of a ship, then the opposing player informs them that they managed to hit part of a ship, but no other information is given. There are two ways to place the ships on the board (horizontally and vertically). So, a player has to first figure out where a ship is, by getting an initial hit, and then guess where the rest of the ship lies. By trying out more points on the board, the uncertainty around the location of the ships is reduced round by round.

Think of the famous 1976 "Four candles" sketch in *The Two Ronnies*. Asking the shopkeeper for "fork handles" is phonically the same as asking for "four candles;" it's a homophone. If you're in a hardware shop that sells both, there's a 50 percent chance that the customer wants candles. But if you're in a candle shop, that's new information, and the probability that the customer is asking for candles—not fork handles—increases.

The main concept behind Bayes' theory was the following:

Initial belief + new information = New belief

In modern terminology, this becomes:

Prior + Likelihood = Posterior

You'll learn more about this next.

Bayes never focused on the theorem that now bears his name. The mathematicians of his time were unhappy with his approach, for two reasons. First, "guessing" didn't sound too rigorous of a process. Secondly, in the absence of information, Bayes assumed that all outcomes are equally likely. In modern statistical parlance, we would say that the prior is uniform. Having to assign a prior probability of belief seemed like an additional hurdle.

The theorem was independently landed on by one of the most prominent mathematicians of all time: Pierre-Simon de Laplace. Laplace rediscovered this principle and published it in 1774. In 1781, Laplace eventually learned of Bayes' earlier discovery, when Laplace visited Paris. Laplace improved his formulation and decided to test it out.

Laplace wanted to know if the fact that more boys than girls were being born was a law of nature, or just a statistical anomaly. He collected records from London, Paris, St. Petersburg, and rural areas in France, Egypt, and Central America. Using his theorem, he managed to conclude that indeed it seemed to be a law of nature.[2]

Laplace went on to make major contributions to other scientific fields such as astronomy. There is a famous expression attributed to Laplace. The story goes that Napoleon asked him why he didn't include God in his explanations of the movement of celestial objects, to which Laplace answered: *Je n'avais pas besoin de cette hypothèse-là,* or "I had no need of that hypothesis."

Laplace was part of the era of scholars, alongside Newton, who instigated a scientific revolution, with mathematics and reason becoming the main tools to explain uncertainty (in place of religious and metaphysical explanations).

But while Laplace did most of the work on Bayes' theorem, his name was never attached to it.

Laplace later on discovered the "central limit theorem," one of the most powerful and significant findings in modern mathematics (which you'll revisit later). Upon discovering this, he realized that once we possessed large amounts of data, the Bayesian approach converged to the traditional frequentist approach. And so, Laplace converted to frequentism, which he abided to until the end of his life.

It's important to note that while the main argument these days is between the frequentists and the Bayesians, there are some other opinions about the subject of probability worth discussing.

The classical definition of probability, used by Laplace, Jacob Bernoulli (the late 17th century Swiss mathematician), and Blaise Pascal (the 17th century French mathematician known for "Pascal's wager"), used the concept of a sample space: In order to get the probability of an outcome, we simply divide the total number of favorable outcomes by the total number of outcomes. Bernoulli also came up with what is called the "principle of indifference." According to the principle of indifference, if we're studying a number of outcomes and those outcomes are indistinguishable from each other, except for their name, then they should be assigned equal probabilities. This principle and the classical definition lead us to assign the probability of rolling 1/6 on each side of a die.

---

[2] We now know the ratio to be approximately 105 boys to 100 girls: `www.searo.who.int/entity/health_situation_trends/data/chi/sex-ratio/en/`

The view of an event converging to its true probability at the limit of infinite trials came later, and the people who influenced it the most were Ronald Fisher (1890–1962), von Mises (1883–1973) and John Venn (1834–1923). This definition is more closely associated with frequentism, and because of this concept of trials, it can't assign probabilities to one-off unrepeatable events, like elections and football matches.

Charles Peirce (1839–1914) and Karl Popper (1902–1994) both supported the propensity view of probability. That is, certain events have a *propensity* to occur under particular conditions. A coin has a propensity of flipping heads 50 percent of the time. This is due to physical factors that give it this propensity. The law of large numbers plays a central role in this definition (you will learn more about this law later). According to this view, after a large number of trials, we will get heads 50 percent of the time, because this is a manifestation of the propensity of the system.

Finally, there are two views on probability that more closely correspond to the Bayesian view. Bruno de Finetti (1906–1985) came up with a view of probability as subjective belief, which Frank Ramsey (1903–1930) also supported. Jeffreys and Rudolf Carnap saw probabilities as an extension of classical logic. The orthodox definition of probability belongs to the Russian mathematician Andrey Kolmogorov (1903–1987), who provided the axioms required to define it

As all of these definitions have now either been pooled under the frequentist or the Bayesian view, let's continue our exploration of these two theories.

The Formulation of Bayes' Theorem

Don't be put off by the esoteric appearance of the following equations. At first glance, you might feel like you've been sent back to an incoherent algebra class, but understanding Bayes' theorem is likely to be easier than you think.

Bayes' theorem is formulated as follows:

$$P(B) = \frac{P(A)P(A)}{P(B)}$$

Recall that Bayes used conditional probability. This is the probability of an event given another event. The conditional probability of an event A taking place, given an event B, is written as P(A|B).

Bayes' Theorem translates in plain language as follows:

> "The probability of A taking place, given that B has taken place, is equal to the probability of B taking place, given A has taken place, times the probability of A, over the probability of B."

Another way to break the denominator down (which might make things clearer for those with more mathematical affinity) is the following:

$$P(A \mid B) = \frac{P(B \mid A)P(A)}{\sum P(B \mid A')P(A')}$$

The probability of B taking place is the sum of the conditional probability of B, given any possible A. So, let's consider an example. Let's say that you are aware that in your town, some people are avid drinkers of the French martini. Those who like French martinis also like smoking. Let's define A=French Martini Lover (FML) and B=Smoking. The Bayes 'expression turns into the following:

$$P(Smoking) = \frac{P(FML)P(FML)}{P(FML)P(FML) + P(\neg FML)P(\neg FML)}$$

The denominator gives us the total probability of someone smoking, regardless of whether they like French martinis or not.

Figure 4-3 shows this probability in a Venn diagram. Given that there are people who are both smokers and lovers of the French martini cocktail, the probability of someone being a French martini lover is affected by whether they are a smoker. So, if I asked you what is the probably of someone liking a French martini when they are a smoker, you would simply divide the area in Figure 4-3 that occupies Both with the area of Smokers.

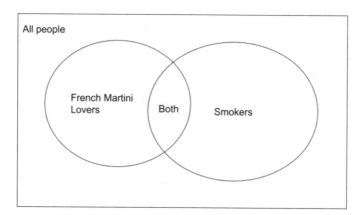

Figure 4-3. *Venn diagram that represents an example universe in which the previous formula resides*

French martinis aside, Bayes' theorem has numerous uses. First of all, Bayes' theorem is used in its raw form in many problems. Second, it has given rise to Bayesian statistics and other models (such as Bayesian networks) with multitudes of uses.

One of the most useful examples of Bayes' theorem is its use in medical tests. Let's say that you suspect you might suffer from a disease that affects one out of 1,000 people. You decide to go to the doctor to do a test. The test has accuracy 99 percent of the time. You do the test, and it ends up being positive. What is the probability you have the disease?

Many people would say that they are certain they are sick. However, the actual probability of having the disease in this case is around 9 percent. Which is not that high. However, if you do two tests, and both are positive, then the probability jumps to 90 percent.

It becomes apparent that Bayes' theorem has a use in medical statistics. But this is only one of the multitude of applications it has. Bayes' theorem allows us to include prior knowledge in our calculations. The values assigned might be educated guesses and are imprecise, but they provide a starting point and at least a rough basis for comparison and analysis.

Bayes' theorem also plays a prominent role in other models. Bayesian networks are a popular class of model that lets us express interdependencies between different variables. The most famous and simple example of a Bayesian network is shown in Figure 4-4.

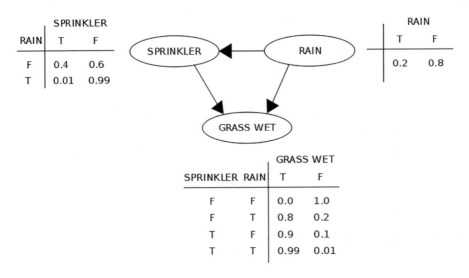

Figure 4-4. *Example of a Bayesian network*

The circles (called "nodes") represent variables, statements about the world that can be true or false. By observing one variable, we can learn something about the rest of the variables, depending on how they are connected. The arrows demonstrate the dependencies. For example, if the grass is wet, then this is either due to rain or due to the sprinkler being on. However, if it rains, the sprinkler will, most likely, not turn on. If we observe that the grass is wet, what is the probability that it rained? And how does this probability change if the sprinkler was also on? While this example might seem very simple, more complicated versions of this model can be used for problems like diagnosing illness.

While this example is simple, there are more complicated versions of networks in which all causes and effects are clearly drawn out. At the time of writing, a search for Bayesian network on Google scholar provides more than 1.8 million results.

While Bayes' theorem is very useful, it was not always accepted in the academic mainstream. It mostly lurked in the background, until recently it became popular again as a very useful tool in modeling uncertainty. Let's look at a brief version of its history.

After Laplace

Before Bayes' resurgence, the "central limit theorem" stole the show, and authorities started collecting data on everything, the opinion that probability could be subjective sounded specious and non-rigorous. John Stuart Mill argued that probability was ignorance, coined into science.

However, some scientists still used it. Once such person was Joseph Louis François Bertrand (1822–1900), a French mathematician with contributions to many scientific fields, from economics to thermodynamics. Bertrand devised a method, based on Bayes, for artillery firings. The method dealt with uncertainties, such as the wind, or the enemy's position, in order to optimize the firings.

The French polymath Henri Poincaré (1854–1912) intervened in the famous Dreyfus Affair, proffering Bayes' theorem in court as evidence that Lieutenant Alfred Dreyfus wasn't a traitor. It was one of the most famous trials in modern French history and Bayes' theorem saved Dreyfus from life imprisonment in what was then French Guinea. Bayes' theorem is arguably the only sensible way to treat evidence in court, and Poincaré recognized this, in spite of him being a frequentist.

It is deeply unfortunate that a court in England in 2011 ruled that Bayes' theorem can no longer be used in trials.[3] It might seem odd for a mathematical theory (or lack thereof) to have such a bearing on—for example—cases of first-degree murder. But that's exactly what happened to Sally Clark in 1999, when she was wrongfully convicted of smothering her two infant sons; the jurors and judges erroneously rested their decision on the statistical unlikelihood of two siblings dying of cot death. In fact, it was far more statistically rare for a mother to willfully kill both her children, and Clark's sentence was overturned and she was finally freed in 2003. It might sometimes be difficult to grasp the intuition behind it, but Bayes theorem is very powerful.

Another prominent statistician we should acknowledge is Sir Harold Jeffreys (1891–1989). Jeffreys and the physicist Bertha Swirles (1903–1999) wrote a book simply called *Theory of Probability* in 1939. Their book was, for many decades, the only book explaining how to practice Bayesian statistics in science. Jeffrey's contribution to Bayesianism also includes the famous "Jeffrey's prior." Jeffrey's prior is used when we are ignorant about the prior distribution.

The great mathematician and inventor of computer science, Alan Turing (1912–1954) also used Bayes in order to crack the Enigma machine. The Enigma was the encryption code used by the German forces during the Second World War. In the process of this, he devised a concept called "ban," whose definition is very similar to our current concept of a "bit" in computer science. Indeed, Claude Shannon (1916–2001), the father of information theory, used Bayes in his development of the theory. He realized that uncertainty can be quantified, and that information reduces uncertainty, whereas encryption increases it.

---

[3] theguardian.com/law/2011/oct/02/formula-justice-bayes-theorem-miscarriage

Russian mathematician Andrey Kolmogorov (1903–1987) discovered that probability theory can be derived from basic mathematical axioms. He suggested the use of the Bayesian method developed by Bertrand for firing artillery, when the Germans attacked Russia in the Second World War.

Other contributors to Bayesian statistics during the 20th century include IJ Good (1916–2009), who was one of Turing's assistants, and Dennis Lindley and LJ Savage. Good wrote the book *Probability and the Weighing of Evidence* in 1950. Dennis Lindley (1923–2013) managed to open around ten Bayesian statistics departments in the UK. Savage (1917–1971) published *Foundations of Statistics* in 1954.

Despite these seminal—indeed, even history-altering— contributions, frequentism dominated the landscape of much of the statistics of the 20th century. The most prominent names of this approach are Ronald Fisher, Jerzy Neyman, and Karl Pearson. Ronald Fisher (1890—1962) was probably the most influential of them all—he came up with many of the theorems and methods that have defined modern statistics, including the "maximum likelihood" and "design of experiments." Neyman and Pearson discovered the "Neyman-Pearson lemma," which plays a huge role in hypothesis testing.

Hypothesis testing and linear regression became the main tools employed by various different sciences. The frequentist approach to those methods became, and still is, the dominant one in scientific publications and in government research bodies. In general, frequentist methods work well when we can structure experiments, or when the repeated experiment analogy makes sense. So, in fields like gambling, insurance, crops, and genetics, frequentist methods might be the best choice. But when we don't have enough information, they break down.

When we know that an event is possible, but we have no data on it, then the frequentist interpretation of statistics cannot help us. Similarly, it cannot help us with unique events, often those that have potentially colossal impacts like elections—or pandemics.

Robert O. Schlaifer (1914–1994) and Howard Raiffa (1924–2016) worked on the problem of decision making in business. Executives are often short of data, and many decisions have to be made based on subjective opinion. This is where the Bayesian approach to probability lends itself naturally. In 1959, Schlaifer wrote the book *Probability and Statistics for Business Decisions*, and then in 1961, Schlaifer and Raiffa wrote the book *Applied Statistical Decision Theory*. Both of these books are standard reads in business literature. They called their approach "decision making under uncertainty," now called Bayesian decision theory.

John Tukey (1915–2000), a prominent frequentist statistician and often credited with coining the term "bit," is also rumored to have used Bayesian methods in order to predict elections. He successfully predicted the 1960 election between Nixon and Kennedy, and he went on to work for two decades with NBC. However, he never publicly disclosed his methods. It was only in 2008, when statistician Nate Silver[4] used Bayesian methods to correctly predict the outcome of the U.S. elections, that Bayesianism gained prominence in election forecasting.

However, while the success of Silver's predictions for the American election of 2008 had been attributed to Bayesian statistics, not many proponents of Bayesian statistics were as excited when, in 2016, he completely missed the mark. We explore this further in the section about forecasting.

There are two more historical examples that demonstrate the usefulness of Bayesian statistics when studying unique events. Norman Rasmussen (1927–2003) was commissioned by the U.S. Atomic Energy Commission to study the probability of a nuclear accident. In 1975, he published WASH-1400, *The Reactor Safety Study*. He used a Bayesian approach, but the subjective view of probability was still unpopular at that time. The U.S. Nuclear Regulatory Commission withdrew its support for the study five years later, and then two months after that, the nuclear accident at Three Miles Island happened. When a contractor was employed to study the probability of a shuttle failure, they estimated (using Bayesian methods) the odds to be one in 35 of an accident taking place. NASA's estimate was one in 100,000. A few years later, in 1986, the Challenger shuttle exploded, killing all seven members of the crew.

In the last two to three decades, Bayesian statistics have witnessed a resurgence. The first reason is that the increase in computational power and Markov Chain Monte Carlo methods have made it possible to actually *compute* Bayesian models. Historically, Bayesian models were even more difficult to compute than traditional statistical models. It was impossible to develop more advanced models until after the 1980s.

Another reason that Bayesian statistics have witnessed a revival is their overlap with machine learning. In fact, a part of the theory behind machine learning has been based on Bayesian probability. You'll learn more about machine learning later, but, for example, Kevin Murphy's textbook *Machine Learning: A Probabilistic Perspective* is a good example of this approach.

---

[4] Nate Silver correctly called 49 out of 50 states. He runs a popular blog at fivethirtyeight.com/.

There are proponents of Bayesianism that believe that any data-related algorithm (including machine learning algorithms) should be based on Bayesian statistics. For example, neural networks, one of the most successful algorithms in machine learning, also have a Bayesian version. However, the debate is still open, and so I take no firm position on the matter in this book. It's up to you to decide—to Bayes or not to Bayes?

CHAPTER 5

What's Math Got To Do With It? The Power of Probability Distributions

"Probability theory is nothing but common sense reduced to calculation."

—Pierre-Simon Laplace

A "probability distribution" is one of the most significant concepts ever devised in mathematics. In an uncertain context, we can't predict the outcome 100 percent of the time. If we're trying to predict the next roll of a die, or the outcome of a sports game, we will get it right sometimes, but not all the time. What if we had a way to express the probabilities of the different outcomes? From there, we can then calculate useful results, like the most likely outcome, or other useful quantities like the variance (which you'll look at shortly).

Chapter 1 explained that the two kinds of uncertainty are epistemic and aleatoric. Epistemic uncertainty exists because of a lack of knowledge. For example, let's say that we are using a medical instrument to take some blood measurements. We want to use these measurements to predict whether someone has contracted a serious virus. These measurements are not 100 percent accurate. In fact, we know that any measurement could be around 20 percent off. In ten years' time we might have developed an instrument that is more precise—say only 10 percent off. This is epistemic uncertainty.

S. Kampakis, *Predicting the Unknown*, https://doi.org/10.1007/978-1-4842-9505-2_5

In a coin-tossing experiment, we face aleatoric uncertainty. Assuming the coin is fair, whether it lands on heads or tails is a random phenomenon. Likewise, we can't know what the result will be before someone rolls the die. The word "epistemic" comes from the Greek for "knowledge," *episteme*; "aleatoric" comes from the Latin word for a die, *alea*—or *aleator*, a dice player. It's a useful way to remember the distinction.

Probability distributions give us a concise way to express both kinds of uncertainty. Most probability distributions are not arbitrary constructs, but emerge from certain models of real-world processes.

Let's consider an example that studies the mean height of males in the United States.

Would you expect to see someone with a height of 180cm (5'10")? This is a reasonable height. Would you expect to see an adult with a height of 250cm (8'2")? That's pretty unlikely. Similarly, not many adult men are 150cm (4'9") tall. The intuition here is that some values are more likely than others. We want to capture this, but also determine *how much* more likely some values are than others.

If you collect a sample of the population with a total size of 1,000, you might end up with a histogram like the one shown in Figure 5-1.

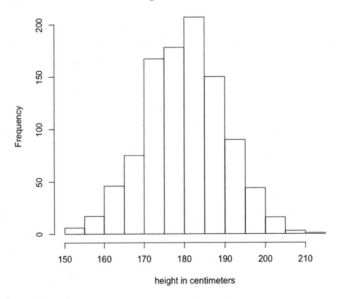

Figure 5-1. *Histogram of 1,000 measurements of height in centimeters*

It is a well-known fact that the height of humans can be modelled pretty accurately through a Gaussian distribution. The Gaussian distribution is named after Carl Friedrich Gauss (1777–1855), the 18th century German mathematician. It shows up in many physical phenomena, as well as in important theorems in statistics—such as the central limit theorem, about which you'll learn about later.

Why is this an important result? The normal distribution is characterized by some certain properties. First of all, it has two variables, which we call parameters, and are used to completely characterize the distribution. These are the mean (symbolized by the Greek letter μ) and the standard deviation (symbolized by the Greek letter σ), or the variance (which is the square of the standard deviation). The mean is the center of the distribution. The standard deviation describes how much the distribution is spread out. See Figure 5-2.

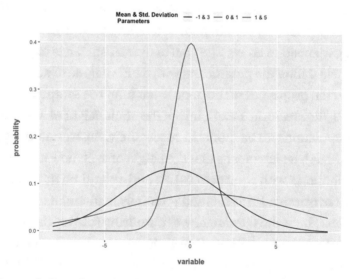

Figure 5-2. *Normal distributions for different means and standard deviations*

So, why is all this important? It is useful for two reasons:

1) It captures our intuition of which values are more likely.

2) The normal distribution provides assurances as to the proportions of people who fall within certain brackets.

The normal distribution has the following property: 95 percent of the subjects are within two standard deviations away from the mean, and 99 percent are within three standard deviations from the mean. While we can't be sure of the exact value of the next sample, we can have degrees of confidence as to where this value will lie.

Let's go back to the example of studying the height of men in America. The mean of this histogram is 180 and the standard deviation is 10. That means that if we choose a random individual from the population, 95 times out of 100, this individual will be between 160 and 200 centimeters tall. If we collect data from the full population, then we would have the true mean of the population, which might be, let's say, 181.23 centimeters, with a true standard deviation of 10.54. In this case, we have completely eliminated any epistemic uncertainty from our estimates.

However, we still face aleatoric uncertainty over the height of newborn males, with each new member who arrives in the population. This problem can be solved through a statistical framework by using a model of Gaussian distribution for the population.

Earlier it was mentioned that the Gaussian distribution is fully characterized by two parameters: the mean and the standard deviation. This is an important concept in parametric statistics.[1] If a phenomenon follows a certain distribution, then this distribution is characterized by certain parameters, and knowing these parameters tells us everything we need to know. The goal of the statistician is to get a large enough sample that will help define the parameters with a high degree of accuracy. Once this is done, we can infer the properties of the population from the sample.

So, in this case, the aleatoric uncertainty of the adult height of a newborn member of our population can be modeled successfully. So, we could say that $height \sim N(181.23, 10.54)$, which means that the height of each new member is modelled by a normal distribution,[2] with a mean of 181.23 and a standard deviation of 10.54.

As you might have noticed, this method is entirely mathematical—we haven't relied on subjective beliefs about the parameter we're studying (the height of males) at any point. It is a purely frequentist method. One of the reasons that frequentism has been so prominent is partly because of the central limit theorem—and related theorems—which, without any mention of subjective beliefs, do a good job of explaining how we can reach the true value of a parameter.

---

[1] Parametric statistics assumes that the data is drawn from a population that can be modeled adequately through a probability distribution. Non-parametric statistics, on the other hand, makes looser assumptions about the population. Which method is best largely depends on the specifics of the case being studied. The curious reader will find countless resources on the subject, but, in brief, parametric statistics are preferred when the assumptions are being met, as their methods can be more powerful. However, in many cases, non-parametric statistics is the only solution.

[2] This is just another name for the Gaussian distribution.

Frequentism has a certain intellectual appeal. We can reach a state of truth about the world, without having to resort to supposition. It also better matches the persistently popular romantic ideal of the scientist being someone who can reach objective truth through study and process. It offers, in essence, *certainty*.

But Bayesianism has its own interpretation of these processes.

In frequentism, we believe that the parameter we are looking for is fixed, and the data is random. An important concept is that of a "confidence interval." We saw an example of this concept earlier on, with the normal distribution. A confidence interval under a frequentist interpretation is a range of values. Within this range of values, we expect to see the true value of the parameter a certain number of times. Therefore, if we are working with 95 percent confidence intervals, we expect that 95 percent of the time, the true value of the parameter will be within this range of our sample.

In Bayesianism, the probability is a measure of uncertainty around values, so we take the data as fixed and the parameter as a random value. Instead of a confidence interval, we use a credible interval. A 95 percent credible interval means that there is a 95 percent probability that the true value of the parameter lies within this interval.[3]

This difference might seem esoteric, and not particularly interesting to someone who is not a statistician, but as we saw earlier, the very fact that we now have an objective language by which to discuss uncertainty is of huge value. How else can we talk about and ascribe value to uncertain propositions? It would be like trying to describe an orange without knowing the word "orange."

There is a heated debate about which interpretation is better. The frequentist approach has dominated statistics for most of its history. However, Bayesian methods have been incredibly successful in some problems. Usually, this takes place when two conditions are met:

1) We have prior knowledge about a problem.

2) We have limited data; for example, we have no data about events that we know could happen in principle.

---

[3] Going back to the example of studying the height of males in America, let's say that we collect a sample of 1,000 men, and we get the 95 percent interval [1.52, 1.94]. According to the frequentist interpretation, there is a 95 percent probability that the true value lies in there. Let's say that we collect a new sample, of 1,000 more men, and this time we get a 95 percent confidence interval of [1.53, 1.92]. Again, there is 95 percent probability that the true value of the parameter will be in there. According to a Bayesian theorist, the interpretation is that, with 95 percent probability, the true value will be between this range.

Prior knowledge refers to things we know about the world. For example, if we are using a statistical model to predict sports outcomes, maybe we know that team A is better than team B. Bayesian methods can handle this very easily. Also, there are some events which are semi-unique (we talk about this later). Elections, for example, happen at semi-regular intervals in all democratic countries, but each time, many things can be different (the candidates, for example). Contrast this to rolling dice, where each roll is more or less the same, and we can repeat this experiment an infinite number of times. Baysian methods are a better choice for events that are semi-unique.

Other Distribution Models

There are myriad distributions to account for all sorts of problems and uncertainties. We will explain a few of these here, but feel free to ignore this part if you find it too technical. Another famous one is the Poisson distribution—so-called after the 19th century French mathematician, Simeon Denis Poisson—which is used to model the occurrence of rare events, like earthquakes. See Figure 5-3.

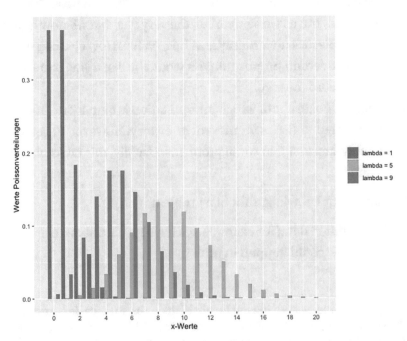

Figure 5-3. *The Poisson distribution for different values of the parameter lambda*

Another popular distribution is the exponential distribution. This is a distribution of waiting times. If we are modeling the arrival of ambulances, we might ask how long will it take until the next ambulance arrives? See Figure 5-4.

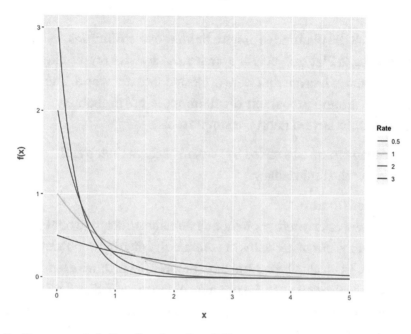

Figure 5-4. *Exponential distribution for different parameter settings*

Then there is the geometric distribution, which can tell us how many failures we will get until we reach a success. For example, how many times do I need to roll a die, until I roll a 6?

The binomial distribution describes the number of successes in a number of trials— for example, if I am betting on getting a 6 on a roll of a die, how many successes will I get if I roll the die 100 times?

Issues with This View of Uncertainty

Probability distributions are a great tool, but there are two main issues that are still hotly contested:

1) What I call the platonic vs pragmatic view of distributions. According to Plato, objects like circles and squares represented ideal forms in heaven that were projected into our world. The question around probability distributions is, are probability distributions laws of nature or simple tools?

2) The second issue was raised by Nassim Nicholas Taleb and concerns "the ludic fallacy."

Let's look at these in turn.

The first dilemma asks whether phenomena in nature seem to obey the laws of certain distributions out of *necessity*, or whether the distributions are a useful approximation. For example, the central limit theorem (which we examine in the next part) provides some intuition as to the ubiquity of the normal distribution in nature.[4] Complexity theorists also tried to find evidence that many phenomena in nature follow certain power law distributions.[5]

These distributions show up in many complex systems in nature, like the distribution of income. When you hear about the 1 percent controlling 90 percent of the world's wealth, that's a power law distribution in effect. Figure 5-5 illustrates this concept.

---

[4] The Gaussian distribution has been used to model all kinds of events and phenomena, from physical attributes (e.g., height), to the noise in measurement instruments. It is found so often in practice, that it looks a bit like a "statistical law of nature."

[5] Power law distributions are called such because they are defined by power relationships of the form $f(x) = kx^{-a}$. They have fat "tails", which means that there is a long range of potential outcomes.

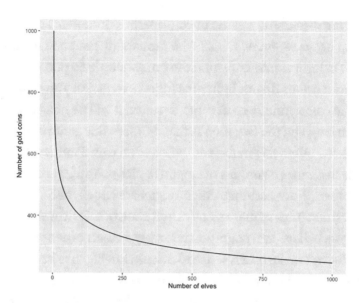

Figure 5-5. *Example of a fat-tailed distribution*

In Figure 5-5, you can see the plot of the wealth distribution of a fictional elven society that uses gold coins as its main currency. You can see that the majority of gold coins are owned by a small number of elves. There are only 23 elves that hold more than 500 gold coins, and one elf that owns 1,000 gold coins. While this example is fictional (or it might be not, depends on whether you believe in elves), the distribution of wealth in modern societies looks very similar.

Another problem with the use of probability distributions is that, most times, they are just used as models of the world. We have to make assumptions and often over-simplify things to account for that—which means that in many cases the models are completely off. Nassim Nicholas Taleb calls this the "ludic fallacy" in his book *The Black Swan*. The term *ludic* is from the Latin *ludere*, meaning play—which in turn comes from the Latin for sport or game, *ludus*. According to Taleb, the ludic fallacy occurs when we believe that a real-world phenomenon can be accurately modeled like a simple game of chance.

To illustrate this Taleb gives the example of two men:

- Dr. John, who is regarded as a man of science and logical thinking.

- Fat Tony, who is regarded as a man who lives by his wits.

The two men are discussing a game of coin-tossing. Someone asks "Imagine that the coin comes up heads 99 times. What is the probability that the next time heads will come up again?" Dr. John would reply that the probability is 50 percent, as the coin has no memory and past events do not influence future events. Fat Tony would reply that it's most likely that the coin is unfair, so in the next round it will be heads again.

The difference between the two men is that one sees this game as an abstract example, whereas the other also perceives the external world influencing the game. Later on, we'll discuss more examples of cognitive biases. You'll see that the orthodox approach to cognitive biases assumes that humans are wrong, because they aren't thinking in the abstract rules of mathematics and statistics. However, other approaches like rational analysis believe that cognition is largely a result of adaptations to the environment: It doesn't make much sense to discuss intelligence in an entirely abstract way. Fat Tony exemplifies human intuition about a real-world scenario. Anyone who is not educated in the laws of probability might look "dumb" in examples like this, as judged by a mathematician. However, a mathematician would look "dumb" in the real world, when the coin ends up being unfair.

Bounds and Limits

"The determination of the average man is not merely a matter of speculative curiosity; it may be of the most important service to the science of man and the social system. It ought necessarily to precede every other inquiry into social physics, since it is, as it were, the basis. The average man, indeed, is in a nation what the center of gravity is in a body; it is by having that central point in view that we arrive at the apprehension of all the phenomena of equilibrium and motion."

—Adolphe Quetelet, 19th century Belgian astronomer and statistician,
A Treatise on Man and the Development of His Faculties

An impressive detail in mathematics is that certain mathematical laws hold true in nature, under all conditions, without requiring any experimental or empirical evidence. In fact, some laws even hold true in all kinds of possible universes, in which the laws of mathematics are the same as ours. Those theorems have then assisted scientists in their quest for knowledge. In the study of uncertainty, there are some theorems that clearly stand out as being unfailing, like the law of large numbers and the central limit theorem.

One of the most powerful theorems in probability theory and statistics is the *law of large numbers (LLN)*. If, for example, we flip a coin a large number of times, we will get a probability of getting heads of around 50 percent, which is the correct theoretical probability.

Poisson referred to the law of large numbers in 1837, but the LLN has an even older history than that. Gerolamo Cardano (1501–1576) was the first mathematician to observe that empirically. He saw that, by conducting a large number of trials, the accuracy of statistics seems to improve. Jacob Bernoulli was the person who first managed to mathematically prove this law in 1713, after 20 years of experimentation. He named it the "Golden Law."

While this might seem intuitively true—perhaps even obvious—it is a significant result. It was certainly *not* necessarily intuitive for people living prior to the Information Age. This theorem provides a powerful link between the theoretical world of probability and the real world.

So, how does this relate to the study of uncertainty? In simplest terms, the rule assures us that as we collect more data from the world, we get closer to the truth. But the inverse is also true. When we know the real theoretical probabilities about a phenomenon, if we can perform enough trials, then the individual trials do not matter—eventually the average value will convert to the true theoretical value underlying the phenomenon. This might seem abstract, but there is a whole industry—valued at over half a trillion dollars worldwide[6]—that is based on this theorem: gambling.

You've probably heard the phrase "the house always wins." This happens because the house has set all games in a way that the expected value is always positive for a casino in the long run.

Let's say that someone proposes the following bet to you. You will do a game of coin flipping, and there is 50 percent chance that you lose or win. If the opposing player wins, then you lose all your money. But if you win, you take all of the player's money. There will be one and only one round of coin flipping, as you either take all the player's money or you lose all your money. You both have exactly the same amount of wealth.

This sounds like a pretty bad proposition. Anything could happen in a single round of this game. You might end up completely broke.

---

[6]www.statista.com/topics/1368/gambling/

However, what if we change a few things about this problem? What if instead of both of you having equal wealth, your opponent has infinite wealth? Also, the rules are such that you don't take all of the opponent's money. You simply get back double the money you invested in. So, if you spend $10 and you win, you get back $20. This sounds good. If you get lucky enough you could become rich, right? You have a 50 percent chance of winning each round, and your opponent has infinite resources, so in theory you could win a lot of money.

Well, unfortunately not. There is a famous theorem called the "gambler's ruin." According to this theorem, a player playing a fair game against an opponent of infinite resources will always go broke after a large number of trials. Casinos in this scenario are the player with infinite resources. The casino doesn't care about any single player. As long as it accumulates more and more games, it is going to win.

But what does this have to do with uncertainty? Gambling is a random phenomenon. We can't be sure about the outcome of a single trial. However, given the right conditions, we can control uncertainty to such an extent that the outcome of a single trial is completely irrelevant.

The gambling example is one of the core themes in the history of our fight against uncertainty. There is not a single method or approach. There are multiple. When mathematicians have attempted to solve it, one of the best approaches they've come up with is that of limits and bounds. What is a limit? Well, you saw one earlier when we talked about the coin-tossing experiment. The result of an individual toss is impossible to predict. However, a large number of tosses has some clearly defined properties. These properties emerge as the coin-tossing experiment converges to the limit of an infinite number of trials. The concept of limit calculations in probability theory is based on the intuition that a large number of repetitions will clear out the white noise and allow useful patterns to emerge.

The concept of bounds is incredibly useful in probability theory. When dealing with a quantity characterized by uncertainty, quite often we can't know what the value of the quantity will look like. However, what we *can* know is some kind of boundary, which the value can't exceed. If I told you to guess the height of my neighbor's daughter, you wouldn't be able to make an accurate prediction, especially if I didn't tell you her age (is she 5 or 45, for example?). But if I asked you to give me an upper boundary for her height, and you told me something like three meters, then you would be 100 percent correct.

One of the oldest theorems that uses bounds is "Chebyshev's inequality." Pafnuty Lvovich Chebyshev (1821–1894) was a Russian mathematician and is considered one of the founding fathers of Russian mathematics. He was also the teacher of the famous mathematician Andrey Markov. The Chebyshev inequality theory provides what we call an upper bound to the probability of a random variable. Bounds in mathematics are like upper and lower thresholds. An upper bound acts as a ceiling, and a lower bound acts as a floor. A value cannot be higher than the upper bound or lower than the lower bound. Bounds are a method to control uncertainty by telling us what the plausible values that we can expect are—and demarcating the values that won't crop up. Of course, knowing the *most likely* value would be more useful, but sometimes we have to go only with what we're provided with.

Coming back to Chebyshev's inequality, let's take a simple example. Let's say that we are studying the average income in a country of 100 million people with an average income of $45,000 and standard deviation of $20,000. What is the probability that we observe someone with an income greater than $100,000? Using Chebyshev's inequality, for this example, we would find that the probability would be at most 20 percent.

Chebyshev's inequality is used to prove the weak law of large numbers. If you have access to knowledge about the mean and the standard deviation of a population, you can get an idea of how likely you are to observe very high or very low values. This holds true for any kind of problem in nature, and you don't need to have any other knowledge about the problem or the data.

It's a prime example of the kinds of ways that mathematicians have dealt with uncertainty. Given limited information about something, how much extra knowledge can you extract? How much can you reduce uncertainty from any crumbs of information that you can eke out from nature? Discovering bounds is one way to do it, and Chebyshev was one of the first mathematicians to develop this approach. There are other mathematicians after Chebyshev who came up with theorems for bounds for all sorts of different problems.

Arguably, the most famous and impactful theorem is the central limit theorem, which I mentioned briefly earlier. The first person to touch upon the subject was the French mathematician, Abraham de Moivre (1667–1754). He used the normal distribution to find the number of heads resulting from multiple tosses of a coin. Unfortunately, his concept never took off, and it was quickly forgotten.

The original formulation of the central limit theorem is attributed to Laplace. Part of Laplace's work revolved around astronomy. For example, one of the problems he was trying to solve was calculating the distribution of the sum of meteor inclination angles.

He faced the challenge of the deviation of his measurements from the theoretical mean that he was trying to capture. Records of this work exist already in 1776. Laplace was able to prove this theorem in 1810. However, the latest proof of the central limit theorem is down to the Russian mathematician Aleksandr Lyapunov (1857–1918).

So, what is the central limit theorem? The theorem states that the sampling distribution of the mean follows a normal distribution, with mean equal to the population μ and variance $\frac{\sigma}{\sqrt{n}}$. This might (understandably) seem a little esoteric to someone who is not a statistician, but it is one of the most useful laws of statistics; let's see why.

The first thing you need to understand is the concept of a sampling distribution. Let's say that you are interested in a quantity. For example, you care about finding out the mean height of all men in England. Or you care about finding out the average income in a city. Or you care about finding the average number of zebras.

If you were a person of unlimited resources, you would be able to go and find every individual subject in your research remit and then calculate the average. If you had done this, then you would have the true population mean, μ. But it's unlikely that you have unlimited resources—who does? Plus, the whole premise of statistics is that you can learn something about a population by taking only a *sample*. Instead of reaching out to every person in England, or every zebra in the world, you simply collect a sample of 1,000 subjects.

So far so good. Let's say that there is some other zebra or height-of-men-in-England enthusiast out there who conducts exactly the same research. This person also collects a sample of 1,000 subjects. Would the two samples look exactly the same? Well, no. The samples will be somewhat different.

Let's focus on the zebra example—say one sample says that the average number of stripes per zebra is 24, and the other one says it is 28. Which one is right? Well, both of them are right and wrong at the same time. When we're collecting a sample from a population, we know that we will get different results every time. The true value of the population parameter lies somewhere in there, but we can't be 100 percent certain of its exact point. And when we can't be certain about something in statistics, we prefer to use a probability distribution—enter the central limit theorem.

The values of 24 and 28 are sample means. They are not the population means. However, we hope to infer the population mean from the sample mean. The central limit theorem is the tool that helps us do that. Let's say that 100 different researchers across the world replicate this zebra research. Each researcher took a sample of 1,000 zebras and calculated the sample mean. What would their results look like? You can see that in Figure 5-6.

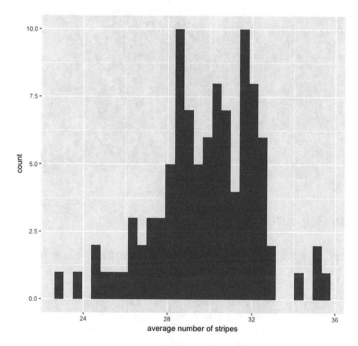

Figure 5-6. *Sampling distribution of the mean number of stripes on zebras*

From this histogram, we can see that some researchers ended up getting means that are below 24, and others got means higher than 34. What happens if this research is carried out by 1,000 researchers? Then the sampling distribution would look like Figure 5-7.

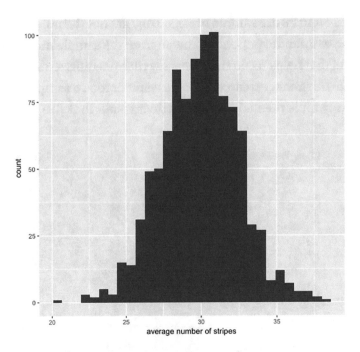

Figure 5-7. *Sampling distribution for 1,000 samples*

And what would happen if we collected 10,000 samples? Then it would look like Figure 5-8.

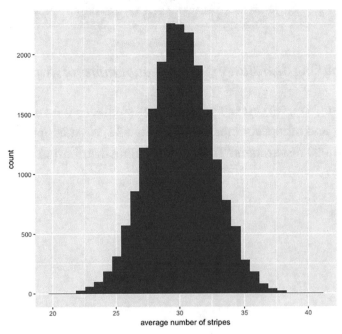

Figure 5-8. *Sampling distribution for 10,000 samples*

Hopefully, you can see a pattern emerging here. The more samples we collect, the more centered this distribution becomes. The uncertainty evens out, and we get closer to the population mean. The central limit theorem tells us that the variance around our estimate (which is a measure of uncertainty) is inversely proportional to the square of the sample size.

It's worth noting that in using the central limit theorem, the relationship holds regardless of the actual distribution that the underlying phenomenon follows. This is what makes it so powerful and explains why the normal distribution shows up in many contexts in physical sciences and engineering.

For example, let's say that we wanted to get the mean of a single roll of a six-sided die. The six-sided die is the archetypical example of a phenomenon that follows the discrete uniform distribution. The probability of rolling a six is the same as the probability of rolling a one. However, when we conduct multiple rolls, the sampling distribution will have a similar shape to the one in Figure 5-8. It will be a normal distribution, not a uniform one.

In Figure 5-9, you can see the distribution of the results of a six-sided die after 10,000 rolls. All results happen roughly an equal number of times.

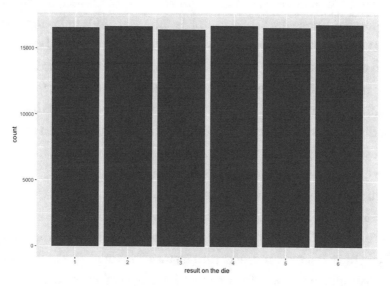

Figure 5-9. *Distribution of the results of a six-sided die after 10,000 rolls*

But when we plot the mean of each sample of results, Figure 5-10 shows what we get. A normal distribution centered on 3.5, which is the true mean of the six-sided die.

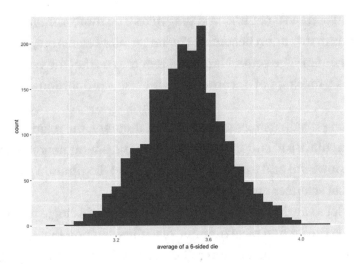

Figure 5-10. *Average of a six-sided die after 10,000 rolls*

As you can see, when it comes to the central limit theorem, zebras and dice have more in common that you might at first have thought.

The central limit theorem occupies a special place in statistics. Many processes in nature are random. Consider measuring the dimensions of flowers, the speed of comets, or the size of animals. These processes can obey different kinds of laws and distributions. But, when we study the average value of this process, we get a normal distribution. This result also explains why the Gaussian distribution shows up in so many natural processes. The reason the Gaussian distribution is also called the normal distribution is because of its ubiquity in nature; the central limit theorem is the explanation behind this.

Secondly, the theorem has played a prominent role in the creation of many other statistical models and hypothesis tests. It makes the mathematical manipulations very convenient—a huge advantage before the advent of the computer.

However, the popularity of the Gaussian as a modeling tool is also one of the reasons that many of our models are wrong. The normal distribution indeed shows up in many places in nature. Also, it is the basis of some very useful models like linear regression. But when all you have is a hammer, everything starts looking like a nail. Models based on the normal distribution are one of the reasons that many of our results are wrong, or our models useless. You'll read more about this a little later on.

No discussion about the normal distribution would be complete without mentioning two scientists who shaped our understanding of statistics: Adolphe Jacques Quetelet (1796–1874), whose quote you saw earlier, and the English statistician, Francis Galton (1822–1911).

Quetelet was a prominent scientist of his time. He helped found the Royal Statistical Society, as well as the International Statistical Congress. He contributed to the creation of an observatory in Belgium, and he also published three books on probability. However, what he is most famous for is his concept of "the average man" (*'l'homme moyen'* in French).

Quetelet conducted a complete census of the French population, and then he measured everything he could—from mortality, to height, weight, strength, insanity, and more. His goal was to study whether the differences between people and places were systematic or random.

In 1835 he published *A Treatise on Man and the Development of His Faculties.* In this book, he provided the vision of "the average man," the person who emerges once we calculate averages of all the attributes that can describe a person. Quetelet was also especially fond of the normal distribution, which he believed was everywhere he looked.

By the current standards of statistical analysis, some of the distributions that Quetelet thought were normal are actually not. The mathematician Francis Ysidro Edgeworth (1845–1926) even coined the term "Quetelismus" to describe the tendency to see normal distributions everywhere.

Quetelet's contribution was that with the measurements of the average man, we can better study deviations from the norm and understand cause and effect. He writes:

> "Above all, we must stop thinking of man in isolation and treat him as merely a fraction of the species. Removing his individuality we will eliminate all that is accidental, and particularities will be erased. [...] This is how we will study the laws concerning the human species, because if we examine them close up it becomes impossible to find them and we get stuck on individual particularities, which are infinite."

Quetelet extended his analysis to other phenomena relating to behavior and the mind, and for that he is largely credited as the father of social sciences. Statistics in social sciences, to this day, are largely concerned with what the average person in a group looks like, how this compares against other groups, and what the causes of those differences are—which is exactly what Quetelet was working to establish.

The concept of the average man fascinated Galton. Sir Francis Galton was an interesting man. He inherited a large fortune, and so he never had to "work" in the way that most people do. He was Charles Darwin's first cousin, and he was an explorer, inventor, and scientist. He was a very curious man, with a thirst for knowledge. He devised a mechanism called the "Gumption-Reviver machine," which would keep him awake between 10pm and 2am by splashing water on this face, so that he could stay up studying and reading. He was also an avid traveler and explorer, having spent a large time in Africa.

Galton believed that "greatness" is hereditary. He noticed that prominent people are usually the progeny of other prominent people. Galton was interested in understanding how talent persists in certain families. He thus came up with a new field of study in 1883, which he termed "eugenics," a discipline later adopted by the Nazis during the Second World War. There is some debate as to whether Galton would have supported the subsequent use of his theory by the Nazis. Galton is still a disputed figure today, since he made many prominent discoveries in statistics, while at the same time condoning some ideas like selective parenthood.

Galton conducted research between the years 1866 and 1869, collecting evidence trying to prove that genius is hereditary, and he published his findings in the book *Hereditary Genius.* According to Galton, "eminence" occurred in a ration of 1 to 4,000 or 5,000 people in England. The people he called "idiots and imbeciles" were around 1 in 400, thus outranking eminent individuals.

Galton also discovered that genius in families does not last for too long. Of the sons of eminent men, only 36 percent were eminent themselves, and then only 9 percent of the grandsons were eminent. Galton sought to explain this pattern, as well as why the data themselves seemed to follow a normal distribution. It was as if the eminent people were above average by the same proportion that "idiots and imbeciles" were below average.

Galton's research would be criticized today for ignoring important social factors, such as wealth and privilege. This omission is not surprising given the social norms of the 19th century. But in spite of that, Galton did discover the concept of "regression to the mean," which proved to be one of the most noteworthy concepts in statistics. This concept describes why short runs of extreme observations will soon revert to looking like an average observation.

A famous example of this is the "Sports Illustrated cover jinx." It's been said that every time an athlete appears on the cover of *Sports Illustrated*, their subsequent performance suffers. This jinx is so famous that it even has its own Wikipedia page.[7]

---

[7] en.wikipedia.org/wiki/Sports_Illustrated_cover_jinx

This is easily explained by regression to the mean. An athlete's performance fluctuates around an average, sometimes being below or above the average. It is entirely possible that an athlete will have a good run of a few amazing games. If this happens, then this athlete is very likely to be featured as the cover star of *Sports Illustrated*. However, it is unlikely that athlete will be able to indefinitely keep up with the standards of their greatest performance. Hence, what is considered a curse is a simple reversion to average performance.

Galton studied this through the use of an ingenious device, depicted in Figure 5-11. He used this in a presentation to the Royal Society in 1874. In this device, pins are dropped, and then they fall into buckets in the bottom. A simple Quincunx, like the one shown on the left, will tend to arrange the pins in a normal fashion. Galton used a machine with two levels. He demonstrated that not only would the pins from the first level array themselves in a normal fashion, but the pins from the second level would do so as well.

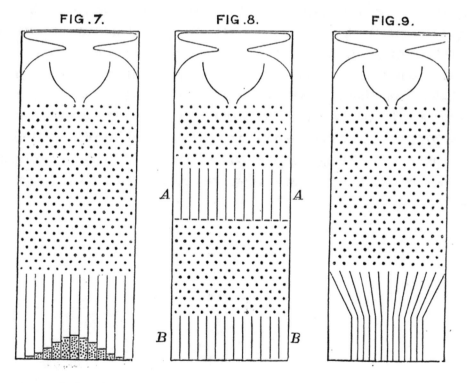

Figure 5-11. *Galton's Quincunx device[8]*

[8] upload.wikimedia.org/wikipedia/commons/7/7f/Quincunx_%28Galton_Box%29_-_Galton_1889_diagram.png

This discovery was monumental, because it showed that even if some observations are extreme on the first level, there will be a tendency for them to revert back to the mean of the normal distribution.

While this example might seem abstract and mechanical, Galton later confirmed this intuition with height data for 928 adults and 208 parents. Heredity ensures that tall parents produce tall children, but regression to the mean ensures that there are no people ten meters tall living among us. This line of work also produced the concept of correlation. Galton is said to have discovered it around 1888.

The concept of regression to the mean explains the regularity of our world, and underlies many of our common-sense attempts to forecast events. We know that a football team can't keep on winning games forever. A stock can't go up forever, and the good times might be followed by bad times, and vice versa.

But when the mean itself has changed, regression to the mean can't help us. There are cases of stocks that crash, with the companies going bankrupt, without ever returning to their previous mean. The world economy has been growing over the decades, so the average size of an economy is always higher. Understanding if the mean has shifted or not is not easy, especially for human-made systems, made up of multiple complex interactions (the economy being one such system). However, this doesn't reduce Galton's contribution to science, nor the usefulness of this idea.

Convergence bound, the law of large numbers, regression to the mean—they all belong to the same family of concepts. In an uncertain world, we don't have to be uncertain about everything. Stochasticity and randomness follow their own laws that produce regularities around us, and they give a sense of order in a seemingly chaotic world.

CHAPTER 6

Alternative Ideas: Fuzzy Logic and Information Theory

"I used to think about it this way—that one-day fuzzy logic would turn out to be one of the most important things to come out of our Electrical Engineering Computer Systems Division at Berkeley. I never dreamed it would become a worldwide phenomenon."

—Lotfi A. Zadeh

Probability is not the only way to quantify uncertainty. Many other researchers and scholars have tried to come up with their own theories of uncertainty that attempt to complement or replace classical probability.

One such theory is the theory of imprecise probabilities. You learned in Chapter 5 how the Bayesian formulation of probability treats probability as something subjective. Someone might not be entirely convinced by this approach. Imprecise probabilities try to solve the problem of defining probabilities when the information received is vague. In this case, we might want to come up with a range of plausible probability scores, instead of a single value.

While this theory is not very popular, it has a long history. The first person who tried to solve this challenge was the English mathematician George Boole (1815–1864). George Boole was the creator of Boolean algebra, which forms the basis of computer circuits. He wrote a book entitled *An Investigation of the Laws of Thought*[1], which, among

---

[1] You can download the book for free at `www.gutenberg.org/ebooks/15114`.

S. Kampakis, *Predicting the Unknown*, https://doi.org/10.1007/978-1-4842-9505-2_6

other things, tried to reconcile probability with logic. The famous British economist John Maynard Keynes (1883–1946), in his book *A Treatise on Probability*, formulated a similar approach to the one that is now called "imprecise probability."

The term "imprecise probability" wasn't actually coined until 1991, by Peter Walley. There is now an organization called SIPTA[2] (The Society for Imprecise Probability: Theories and Applications) that promotes the use and theory of imprecise probabilities. While there are some academics working in this field, imprecise probability is not a hugely popular theory. Perhaps one reason is that more traditional approaches to probability seem to work fine on a variety of problems. Therefore, understanding the formulations of imprecise probability might seem like too much work, without a clear benefit to some researchers. It remains to be seen whether this approach will evolve to capture a larger audience.

Another alternative theory of uncertainty is the "Dempster-Shafer theory." This theory was first introduced by Arthur P. Dempster in 1967, and it was then further developed by his student, Glenn Shafer, in 1976. The Dempster-Shafer theory tries to extend the Bayesian interpretation of subjective probability. The theory formalizes the notion of subjectivity through belief, one of the reasons that it is also known as "the theory of belief functions."

In the Dempster-Shafer theory, two measures are used: belief and plausibility. So, let's say that you receive a new piece of information. The belief is a number between 0 and 1 that represents how much faith you have in this piece of information. The plausibility is actually how likely it is for this information to be true. The theory presents a different way to combine evidence from different inaccurate sources. Using two measures might seem more complicated than necessary, but it has been applied to the problem of sensor fusion: merging information from different sensors together.

This theory never reached the mainstream. The reason is that the rules for combining evidence can sometimes produce counter-intuitive results. Plus, much like with imprecise probabilities, more traditional approaches to probability seem to work well enough, so any new approach should justify its practical applicability. However, there are still scholars researching its applications and there is also a society called BFAS[3] (Belief Functions and Applications Society), established in 2010, that promotes it.

---

[2] www.sipta.org/

[3] bfasociety.org/

One of the most successful alternative theories of uncertainty is *fuzzy logic,* discovered by Lotfi A. Zadeh (1921–2017) in the 1960s. Zadeh was born in Azerbaijan, while it was still part of the Soviet Union. When he was ten years old, the family moved to Tehran in Iran. He traveled to America in 1944 and enrolled in MIT, where he received an MSc degree in electric engineering. He received his PhD in electrical engineering from Columbia University in 1949. He moved to Berkley, California in 1959, where he stayed for the rest of his life.

Zadeh was interested in the problem of linguistic uncertainty. This is something that traditional probability was not really covering. As an example, say that you are talking to a friend and you saw someone who is very tall. Or that you were driving really fast on the motorway. What does very tall or really fast actually mean? Maybe "tall" for you is someone who is 185cm (6 feet) or above. But based on this, would some who is 185.1cm be tall, and someone who is 184.9cm be short? There are no clear boundaries. In fuzzy logic terminology, the boundaries are "fuzzy" instead of "crisp."

Lotfi Zadeh first came up with the concept of fuzzy sets in 1965. A fuzzy set is an extension of a standard set in mathematics, where each object can only partially belong to a set. Think about examples in real life. Maybe you are listening to a song, and you feel like the song could belong to multiple music genres simultaneously. Maybe you eat food from an Asian fusion restaurant, and the meal is something between Korean and Japanese. Or maybe you are dressed in a smart casual way for an event. You could say that you are dressed formally, but not completely formal. At the same time, you are casually dressed, but your clothes are not fully casual. You are somewhere in between.

These are real-world examples of the intuition behind fuzzy logic. You can have objects which *kind of* belong to some category, but also kind of don't. Someone can be somewhat tall, and somewhat not tall. Zadeh himself famously said, when asked about his nationality:

> "The question really isn't whether I'm American, Russian, Iranian,
> Azerbaijani, or anything else. I've been shaped by all these people
> and cultures and I feel quite comfortable among all of them."

His nationality and cultural background was "fuzzy." He didn't identify with a single culture or people.

In 1973, Zadeh presented the theory of fuzzy logic. Fuzzy logic determines the rules of reasoning with fuzzy sets. Much like in probabilities, we want to be able to use sets to perform inference for a wide variety of problems. Zadeh did the same for fuzzy sets.

Fuzzy logic is an extension of traditional propositional logic to include fuzzy categories. By using fuzzy logic, we can form propositions like the following:

"If someone is very tall and very strong, they are likely to be a good basketball player."

Zadeh's achievement was to find a mathematical way to capture linguistic uncertainty. We speak to each other all the time using expressions like this one; using a way to encode this inside a computer was not something that had been achieved before Zadeh. If we wanted to express the same sentence in propositional logic, we would face all kind of issues as to what the meaning of the phrases "very tall" or "very strong" mean.

The math behind fuzzy logic looks similar to probability theory. For example, a concept, like tall, gets a score between 0 and 1. Hence, we can say that someone is 0.7 tall. This is why the two theories were confused so many times, but Zadeh's goal was not so much to replace probability as to cover a different type of uncertainty not yet covered explicitly by probability theory. In 1978, after introducing fuzzy sets and fuzzy logic, Zadeh proposed *possibility theory.*

In possibility theory, a state that is impossible is assigned a possibility of 0, and a state that is totally possible is assigned a possibility score of 1. We are using possibility distributions to describe our knowledge of possible outcomes. Possibility theory uses two measures, possibility and necessity, to reason with incomplete information and describe whether an event might happen.

Fuzzy logic has to be the most popular and successful of Zadeh's theories. It found various applications in control systems in engineering. For example, fuzzy logic is used in the following applications:

- Air conditioning systems[4] to control the temperature and the flow of air.

- Rice cooking devices[5] to control the heating of the rice.

- Washing machines[6] to control the water intake, water temperature, wash time, rinse performance, and spin speed.

---

[4] www.researchgate.net/publication/255569821_Fuzzy_modeling_and_control_of_HVAC_systems_-_A_review

[5] home.howstuffworks.com/rice-cooker2.htm

[6] www.samsung.com/in/support/home-appliances/what-is-fuzzy-logic-in-a-washing-machine/

- Televisions sets to control brightness.[7]

- Vacuum cleaners.[8]

- Automotive engineering on things like ABS breaking and antiskid steering.[9]

Fuzzy logic also became a core part of what was called soft-computing and computational intelligence. Soft-computing is one of the many branches of the study of intelligent systems, which have eventually all come under the umbrella term "data science." The term soft-computing came in contrast to hard-computing. Soft-computing aims to build algorithms that can reason under uncertainty and find inexact solutions to hard tasks, like prediction and forecasting. Hard-computing refers to traditional computing methods that deal with exact problems and algorithms for those problems, such as sorting a list. Computational intelligence was based on the theory that we can build intelligent algorithms by mimicking natural processes. The cornerstones of computational theory were evolutionary computation, neural networks, and fuzzy logic. Neural networks are a computational analogue of the brain, an evolutionary computation of natural selection, and the fuzzy logic of language.

Machine learning is now the dominant approach in academia and has absorbed all useful techniques and methods from related fields. Data science has become the dominant term in commercial settings. You won't hear many people saying they are studying soft-computing or computational intelligence. We talk more about those fields in Chapter 8.

Zadeh received multiple awards and honors throughout his lifetime. Until the end of his life, he stressed the importance of studying *possibility* and not just probability. While Zadeh's theories never truly reached the mainstream—aside from being the title of the 1996 debut album from Super Furry Animals—it is used in multiple devices and problems. Alas, many of them are now being replaced with more modern machine learning algorithms. But that in no way diminishes the impact of fuzzy logic.

Zadeh appreciated the fact that uncertainty has various gradients and can appear in multiple ways in our lives. Something fascinating is that while linguistic uncertainty took a long time to put into formulas, it is something that all of us are now comfortable with.

---

[7] waset.org/publications/10439/a-novel-fuzzy-logic-based-controller-to-adjust-the-brightness-of-the-television-screen-with-respect-to-surrounding-light
[8] www.freepatentsonline.com/5251358.html
[9] link.springer.com/article/10.1007/s12544-015-0179-z

We go about our daily lives using language to communicate all sorts of fuzzy concepts, without having to think too much about it. This is, on the one hand, a testament to the power of our brain to deal with uncertainty. On the other hand, it is also a testament to the fact that fighting uncertainty is not a lost cause—but it is a war that is constantly changing forms.

When a creature is alone, it faces lots of uncertainty around its environment, from what food to eat to which animals to fight off. Humans formed tribes to protect themselves against nature, and in doing so, they also managed to gather information about their environment. However, information is more useful if it can be communicated to other people in the tribe. Hence, our species developed language as a form of communication. Language reduced the uncertainty in communication between people, but it also introduced new kinds of uncertainties. We saw that the meaning of many words and concepts is somewhat fuzzy, that is, uncertain. Furthermore, language allowed the development of culture, which also adds more complexity. Think of all the uncertainty you face about the correct rules of conduct when you are visiting a new country.

Communication, information, and uncertainty are closely related. There was one man who studied the relationship between all three and is the person responsible for much of the world we see around us, from mobile devices, to the Internet. This man was Claude Shannon and he is the subject of the next section.

Information Theory—Measuring Uncertainty

No discussion of probability and uncertainty would be complete without talking about Claude Shannon.

Claude Elwood Shannon (1916–2001) was an American mathematician and electrical engineer. He was a brilliant student, coming up with the digital circuit design theory based on Boolean algebra, when he was only 21 and was still studying at MIT. This is now part of the basic principle governing logical circuits in computers. It has been called the most important master's thesis of the 20th century, as well as the "Magna Carta of the information age." He eventually obtained his PhD from MIT in 1940, and then became a research fellow at the Institute for Advanced Study in Princeton. There, he got in touch with various important scientists of his era, like von Neumann, Einstein, and Kurt Gödel. Shannon later joined Bell Labs in 1941 to work on projects related to the

Second World War, such as cryptography and control systems for anti-aircraft gunnery. Part of his work was used on the system that encrypted the communications between Churchill and Roosevelt.

One of the problems that Shannon was trying to solve, while coming up with his theory, was the problem of information transmission over a channel. If you have a transmitter and a receiver, how can you encode the original message in the smallest possible data structure, while ensuring that the receiver can recover the original message without any loss of information?

After the end of the war, the work he did in Bell Labs was declassified, and so he published a paper in 1949 entitled *A Mathematical Theory of Communication*. This paper signaled the beginning of the information age. In this paper, Shannon laid out the foundations of information theory that permeate all aspects of computer science. This paper was the first time the word "bit" was used.[10]

So, what is information, and how can we define it? Shannon connected information to the element of surprise. If an event is unexpected, but it takes place, then we have received a high degree of information. If an event was totally expected, and it takes place, then the simple confirmation that it took place does not offer much new information. Let's consider an example.

Let's say that someone provides you with a piece of information, for example, "the average temperature last July in the Sahara desert was over 5 degrees Celsius (41 degrees Fahrenheit)." Even if you've never been or read anything about the Sahara desert, you can imagine that it must be a pretty hot place. This piece of information is not very useful to you. Essentially, you didn't learn anything new. It was totally expected. Similarly, if someone told you that there was at least one day of snow during the last winter in Canada, again, this doesn't sound like something out of the ordinary. These are uninformative pieces of information.

However, let's say that someone told you that it snowed in the Sahara desert last July. This is a phenomenon—that might be even considered impossible—and, for sure, totally unexpected. You didn't know this could happen. You have learned something new. You have received an informative piece of information.

---

[10] Shannon attributed the term to the famous statistician John Tukey, who had written a Bell Labs memo in 1947 in which he contracted "binary information digit" to simply "bit."

Shannon used the concept of information entropy to model uncertainty.[11] Information entropy is the average surprise among all the possible outcomes of a given situation. So, let's say that we care about the outcome of a political election, and one party is ahead in the polls with 90 percent of the voters supporting it. The information entropy (and hence the uncertainty) is low. You don't expect to be surprised. However, let's say that there are two political parties, each one getting 50 percent support in the polls. In this case, you are facing a maximum degree of uncertainty around the outcome. Indeed, anything could happen. Hence, the information entropy is at its maximum possible value for this event.

This concept has countless applications in telecommunications, signal processing, and computer science. Language can give us another, intuitive example. Imagine that you receive a noisy voicemail and you are trying to figure out what this message is about. Let's say, that because of the noise, the only words you can hear are words like "A," "the," "and," and so on. These words are not very informative. You expect to hear them many times in a few sentences. There is no surprise there, and hence no information. You can't really guess what the message is about.

However, let's say that you hear the words "alien" and "invasion." These words do not occur that often, and they happen even less often together. You would be right to be very surprised if there was indeed some kind of alien invasion taking place! Hence, these words contain lots of information.

While these are concrete examples, talking about events, and language, Shannon completely abstracted all this and simply talked about "bits." It doesn't matter whether the information you convey is in language, images, electromagnetic waves, or something else. Information is always information, and we can use the principles of information theory to analyze it. Shannon used the example of a random process that produces strings of symbols. If a machine produces only one symbol, then the information entropy is 0, as uncertainty is at its minimum possible point. If it produces lots of different symbols, then the information entropy is going to be a number that represents the minimum number of bits needed to encode the message.

This was a profound discovery that reshaped the way we view the world. Any kind of information can be analyzed through those same principles. Besides the obvious contribution to computer science, electrical engineering, and all related fields, those principles extend to other fields. For example, Shannon published an article in 1951

---

[11] This term was inspired by the entropy concept of thermodynamics, about which you learn about in a later chapter.

entitled *Prediction and Entropy of Printed English.*[12] In this article, he analyzed English from a statistical perspective and proved that the use of whitespace—for example a paragraph break—in written language lowers uncertainty. This provided a justification, in terms of information theory, behind this cultural practice.

The importance of Shannon's contribution to science cannot be overstated. He was the one who introduced the concept of "channel capacity," defined as the maximum number of bits that can be transferred over a channel with arbitrarily good reliability. In everyday life, we simply call this the "bandwidth." This is achieved with error-correcting codes. What is an error-correcting code? Let's consider an example.

Let's say that you want to transmit a message over a noisy radio. Both you (the transmitter) and the other party (the receiver) know that the channel is noisy. One out of three words disappears when talking on the radio. So, how could you intuitively fix that? A simple way is to use a strategy—the error-correcting code—that says "repeat each phrase three times." If you follow this strategy, it will take longer to complete the full communication, but there would be a very small probability that the receiver wouldn't hear all the words. If you wanted to feel even safer, you could simply repeat each phrase ten times. The message would take even longer to transmit, but there would be practically zero probability of missing a word. Shannon formulated this intuition mathematically. These principles now underlie all current communications.

Information theory is also directly related to data compression and machine learning. Data compression is one of the most common techniques in computer science, used in everything from MP3 files to cloud storage.

Let's say that you want to transmit the following message:

> "*Jurassic Park* is the best film of all time! *Jurassic Park* is the best
> film of all time! *Jurassic Park* is the best film of all time!"

Is there a more efficient way to convey your enthusiasm about *Jurassic Park*? The message consists of three identical sentences. Instead of sending the whole message, you could encode this into something like the following expression:

> "*3: *Jurassic Park* is the best film of all time!"

This is a compressed form of the original message. You convey the exact amount of information, but using a more compact, shorter representation, which saves you bits.

---

[12] ieeexplore.ieee.org/abstract/document/6773263

So, how does data compression relate to machine learning? David MacKay (1967–2016) described the relationship between the two in his book *Information Theory, Inference and Learning Algorithms.*[13] Compression is related to predictability. As you will see in Chapter 8, the key concept about learning is generalizing from specific examples to a wider concept. When we do this, we are essentially trying to compress information, accepting that the learning process will also incur some mistakes along the way.

When something is easy to predict, it is also easy to compress. For example, let's say that you are trying to compress a song, where every third chord is C minor. An easy way to compress this would be to simply use a code which in simple words might look like this:

"C minor at positions 3, 5, 8, until the final position at 30."

If you were to predict that every chord is C minor, you would still be correct 1/3 of the time. This simple example shows the analogy between the two.

Information theory is also related to evolution. First of all, all of our genetic information is encoded in our DNA. Secondly, evolution can be seen as a learning game. There is a quantity for each organism, called "fitness." The fitness can be anything that helps survival and procreation: from our high IQ, to the ability of birds to fly, and the ability of an antelope to run fast. Natural selection can be seen as a teacher. Each new generation that survives has children who carry information about which adaptations worked.

John Maynard Smith (1920–2004) argued that the difference between apes and humans was 400,000 bits of information. MacKay argued that this is a small difference, given the size of the genome, and developed a different model. The information theoretic view of evolution also explains why sex exists, instead of most organisms simply multiplying through parthenogenesis. A population with a size of genome G can acquire information at a rate $\sqrt{G}$ faster than a parthenogenic population, and hence it can evolve and adapt faster to its environment. Hence, a simple organism with 1,000 genes (the human genome is around 20,000 genes) would evolve 100 faster through sex than through parthenogenesis.

Another field where information theory is applied is cryptography. Cybersecurity is a trending field in technology right now, and for a valid reason. The more our world depends on the Internet, the more data we transfer, and the more transactions we

---

[13] www.inference.org.uk/mackay/itila/p0.html

perform, the greater the risk of this information falling into the wrong hands. We have already talked about how cryptography was used in World War Two, and the role that the efforts of Alan Turing—alongside Shannon and other prominent scientists—played in it.

Shannon was the first person to prove that unbreakable cryptography was possible. He devised a scheme called the "one-time pad scheme" or the Vernam cypher, after Gilbert Vernam (1890–1960), who invented it. Shannon proved that this kind of encoding scheme was unbreakable. The reason that we don't have unbreakable encryption yet is because this scheme is not practically usable. However, quantum cryptography might be able to solve this in the future.

Shannon's multiple contributions were a direct result of his personality. Shannon dabbled in different scientific fields and problems throughout his lifetime, plus he loved unconventional hobbies—at least for academic circles. He was often seen riding a unicycle and juggling at the same time. Sometimes he would use a pogo stick to go around. He was a chess enthusiast, studied the complexity of chess, and even created a program that could play chess. He liked building things, and among the many devices he created, he built a robotic mouse that could navigate mazes (one of the earliest attempts at artificial intelligence), a machine that could juggle balls, and a trumpet that could breathe fire when it was played.[14] He received more than 20 honorary doctorates and awards throughout his lifetime, but he never really cared for them. He hung the doctoral hoods from a device, which he built himself, that resembled a rotating tie rack.

Shannon unfortunately developed Alzheimer's disease in the 1990s. This was a tragic irony, as the man who created the field of information theory was afflicted by a disease that degrades information channels in the brain. He succumbed to the disease in 2001, at the age of 84.

However, before his death, this one man managed to set the foundations of the information age. He was the first person to attack the problem of uncertainty directly, to quantify it, and digitize it.

How exactly is probability used in practice? The science of statistics, which forms the basis of the next chapter, has been the only discipline—until the discovery of machine learning, that is—that dealt explicitly with the study of uncertainty. While statistics might incite boring school memories, it is, as Karl Pearson said, "the grammar of science." And it's a language worth getting your head around.

---

[14] Jimmy Soni, Rob Goodman, *A Mind at Play: How Claude Shannon Invented the Information Age*, 2017

CHAPTER 7

Statistics: The Oldest Kid on the Block

"There are three kinds of lies: lies, damned lies, and statistics."

—Benjamin Disraeli

Statistics get a bad rap.

They can seem purposefully opaque, random, and tangential, screaming out at you like misleading headlines: "Twenty-two percent of people think that Barack Obama is not American!," "Forty-eight percent of Britons don't wash their hands after using the bathroom!," or "One in three marriages ends in divorce!"

Where do these numbers come from? Who exactly are these "people"? Was every single human in Britain asked whether or not they wash their hands?

It's well worth getting to the bottom of statistics. Every concept and theory discussed so far is intrinsically linked to the discipline of statistics; statistics is one of the most important, but least recognized, disciplines in the world of data science.

In fact, it was the first scientific discipline that systematically dealt with data analysis. However, these days, data science has become the more popular buzzword. Many of the techniques first created by statisticians are now promoted as machine learning techniques. Also, the way of thinking established by statisticians has been somewhat discredited in favor of the more experimental approach favored by data science. The next section takes a step back and explains a few things about the history of the field, and covers the differences of the current approaches to data science.

© Stylianos Kampakis 2023
S. Kampakis, *Predicting the Unknown*, https://doi.org/10.1007/978-1-4842-9505-2_7

Descriptive vs Inferential Statistics

When people talk about statistics, they usually think of things they did in school, like graphs, or sport statistics, like the mean number of points scored. However, when a statistician thinks about statistics, they usually think of inferential statistics.

Inferential statistics is a simple but powerful idea, which is that scientists can learn things about a population by drawing a sample from it. Sampling is a concept we are all familiar with these days. When trying to get the pulse of society, pollsters won't ask every individual in a country, but a small part of the overall population. When we go for a blood test, the doctor will not draw all the blood from our body, only a small portion. And when you are trying to find out whether the soup is too hot, you simply taste a little bit with the tip of your tongue, instead of eating the whole bowl. It's the indicative tip of the iceberg.

Sampling was not always a popular concept. The person who introduced sampling was, quite interestingly, not a mathematician, but a 17th century English haberdasher: John Graunt (1620–1674). Graunt was a merchant of buttons and needles. He was quite a successful businessman, and he made enough money to be able to pursue other interests in his free time. In 1662, he published a small book called *Natural and Political Observations Made Upon the Bills of Mortality*. The book contained archives of deaths in London from 1604 to 1661, along with an interpretation that looked like a first attempt at what is now called survival analysis.

Graunt studied the various causes of death, and he came up with probabilities of death from each cause. Graunt estimated that around 36 percent of children died before the age of six. He also estimated percentages of death from the plague, murder, and other afflictions. Graunt created something very similar to modern life tables: A table of data displaying the survival rate at different ages. According to this table, someone who was 56 years of age, for example, had only a 6 percent probability of survival until the age of 66.

He also used these records to come up with an estimate of the population of London. While it was widely believed at that time that there were around two million people in London, Graunt was able to come up with an estimate of 384,000. He calculated that every year, there were around 13,000 deaths on average. He noted that births were fewer than deaths, on average, so he came up with an annual average of 12,000 births. This indicated that there were around 24,000 women who were of childbearing age.

He assumed that a household (which might also include servants and lodgers) had approximately eight people, and that only half of the households had a woman of childbearing age. Therefore, the total number of households was around 48,000, which leads to the total number of 384,000 mentioned earlier.

The Englishman Edmond Halley (1656–1742) published another seminal work in this area in 1693, which was around 30 years after Graunt's work. Halley is mostly famous for his work in astronomy and the comet that bears his name. However, one of his less well-known contributions was in statistics. Halley had been given life and death data for the city of Breslaw.[1] Using this data, he came up with detailed life tables and probabilities of survival in different ages. However, it took around 100 years before governments and insurance companies started using his work. Before that, annuities were sold based on the life tables of the Roman jurist Ulpian (developed in 225AD) or they were not used at all. For example, the English government had sold annuities in 1540, without taking into account someone's life expectancy.

Laplace was also involved in early attempts to use sampling. He calculated the population of France in 1786 by using the ratio of the population to births during the preceding year. Another notable mention is William Petty (1623–1687). Petty lived in London, where he was an economist, physician, and philosopher. He became friends with Graunt and helped with some aspects of his work, and also came up with his own estimation of life expectancy at birth. Petty estimated it to be 18 years, whereas Gaunt estimated it to be 16.

Hence, the seeds of sampling were sown during the 17th century. It was the merging of those techniques with the techniques from probability theory that instigated the idea of *inferring* properties of the population by investigating just a small part of it.

The modern form of statistics is mostly down to the efforts of scientists from the late 19th and early 20th centuries. The most notable names are William Sealy Gosset, Ronald Fisher, Egon Pearson, and Jerzy Neyman. These men set up the foundations for modern statistics by devising rigorous techniques for testing and modeling. These techniques have been the staple of science ever since.

---

[1] Breslaw now is called Wrozlaw and is part of Poland.

Hypothesis Testing: Significant or Not?

William Gosset was the head brewer at Guinness, the famous stout beer producer. In the late 19th century, Guinness had already become the biggest brewery in the world, and it was looking toward science in order to improve its products—much like companies now might look into the use of data science and AI in order to gain an edge. The main focus of the brewery was to maintain high quality in its beer. Between the years of 1887 and 1914, the quantity of beer produced had doubled, reaching nearly one billion pints.[2] Obviously, increasing quantity presents a challenge for quality. Gosset's job was to find an answer to this problem.

Guinness, like many other kinds of beer, is flavored through the use of hops. Guinness measured the quality of the hops by qualities like the look and the fragrance. The challenge was to measure the quality of a batch of hops by taking only a small sample of them.

Until that time, no one else had dealt with inferential statistics on small samples. The challenges of using a sample of only five hops seemed to be somewhat irrelevant to most statisticians of that time, who worked with considerably larger samples. Gosset successfully solved this problem and then spent one year alongside Egon Pearson (Karl Pearson's son) at University College London, improving the mathematical foundations of his method.

This eventually led to the publication of the now-famous, in the statistical world, *Student's Distribution* (see Figure 7-1). Guinness wouldn't allow Gosset to publish his findings under his real name or mention that these techniques were used by Guinness—presumably the idea that each hop wasn't lovingly hand-picked for brewing didn't quite fit with the brand's image. Gosset had to find a way to anonymize himself and his work, and he came up with the moniker, "Student," from his 1906-1907 notebook on counting yeast cells with a hemocytometer, *The Student's Science Notebook*. And so, in 1908, Gosset published his work in the paper, "The Probable Error of a Mean."[3]

---

[2] pubs.aeaweb.org/doi/pdfplus/10.1257/jep.22.4.199
[3] Student. (1908). "The Probable Error of a Mean." *Biometrika*, 1-25.

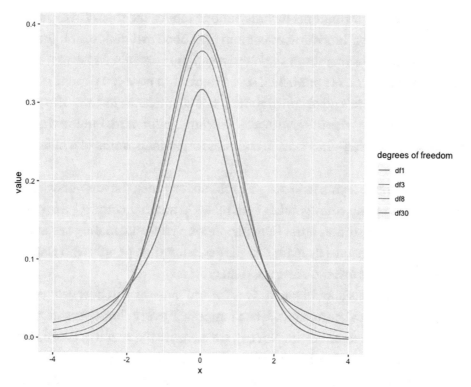

Figure 7-1. *Student's T-distribution for different degrees of freedom. Notice the similarity to the normal distribution*

"Student's" research left a strong impression on a young mathematician named Ronald Fisher. Ronald Fisher was intrigued by student's approach: Using a distribution to estimate how likely something is compared to random chance. Using this concept as a foundation, Ronald Fisher went on to produce an amazing body of work—and in doing so, he became the father of modern statistics.

When Gosset published his now-famous work, Fisher was still studying mathematics at Cambridge. Fisher was born in 1890 into a wealthy family, and he lived in affluent Hampstead in London until he was 15. However, his mother died when he was 14, and his family went bankrupt when he was 15, as the business folded. The Fishers had to move to a small house in the poorer neighborhood of Streatham. Fisher kept studying at the prestigious Harrow School through scholarships.

Fisher suffered from poor eyesight, which was in fact so bad that he was advised against reading under electric light, because it strained his eyes too much. This might have been a blessing in disguise: Fisher developed a remarkable ability to visualize and solve mathematical problems in his head.

Some of Fisher's contributions include, among others, the theory of experimental design, the concept of variance and maximum likelihood estimation, and the analysis of variance (one of the most common statistical tests). He conducted research on biology, agriculture, and eugenics. He published seven books and over 400 papers.

One of Ronald Fisher's great contributions was significance testing. The problem that significance testing tries to solve is relatively straightforward. But the way that it is solved—the solution itself—has caused countless arguments, some of which are still going on.

Imagine that you're trying to answer a problem for which the answer can *only* be true or false. Are males in one country taller than males in another country? Are people who consume more carbohydrates fatter than people who consume mainly fats and protein? Did people who were given a drug for disease get better than people who were left untreated? Are dragons faster on average than griffins?

All of these questions should have a "yes" or "no" answer. But how can we find this answer? For practical purposes, we can't approach every single person in two countries and measure their height. Nor can we measure the speed of every single dragon (they can be pretty hard to get hold of). A potential way to solve this is to resort to sampling: get a few subjects from each category, calculate an average, and then make a comparison.

But how can we make sure that this process has produced reliable results? Let's see how this could go wrong. Let's say that you just landed in a new country. You notice that the first ten people you meet are wearing red trainers, so you assume that everyone in the country must be wearing red trainers. You see the problem with that? This is basically the problem of induction discussed in a previous chapter.

So, there is a similar issue with sampling as there is with induction. Every time we get ten people, dragons, or whatever, our average of whatever quantity we are studying will be slightly different. Just because the average speed of a dragon in your sample is 12km/hour, doesn't mean that this is the average speed of the whole population. Sometimes our sample will give a higher average, sometimes a larger average. In some comparisons, dragons might be faster than griffins, in other comparisons it might be the other way around, or the speed of the two magical animals might be equal to each other.

Hence, we need a method, a scientific process, through which we can infer from our sample—with some degree of certainty—the objective state of the world. Fisher came up with this method. In 1925 he published the book *Statistical Methods for Research Workers,* in which he outlined his method and popularized the (much disputed) p-value—more on that later.

Let's say, for a moment, that we assume that dragons and griffins have equal speeds. We collect two samples, and then we measure the average speed for each one. If they indeed have equal speed, then what would be the probability of observing the same average speed for the two samples?

This might sound a bit mind-boggling, but bear with me. Imagine, for a moment, that the distributions of speeds for the two creatures looked like the plot in Figure 7-2.

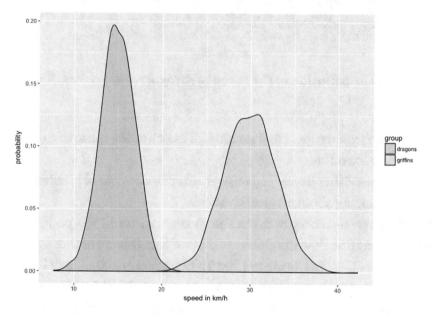

Figure 7-2. *True distribution of the speed of dragons and griffins*

From this plot it looks like the two magical animals have very different speeds. There is a small area over which the two distributions overlap—see the tiny intersection around the 20km/h mark. Hence, there would be a very small probability that we would draw two samples at random and the speed of those samples would be equal.

On the other hand, let's say that the true population distributions looked like the plot in Figure 7-3.

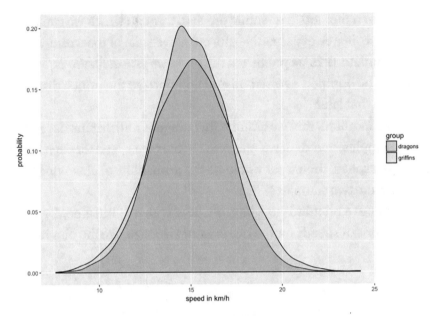

Figure 7-3. *True distribution of the speed of dragons and griffins. Equal speeds scenario*

In this case, we would expect that for most of the samples, the average speeds would be pretty close to each other.

The method that Fisher developed answers exactly this type of problem in a scientific way—even when talking about magical creatures.

Besides Fisher, we also have to thank Karl Pearson who, in 1900, published a paper with a rather lengthy title: "On the Criterion that a Given System of Deviations from the Probable in the Case of a Correlated System of Variables Is Such that it Can be Reasonably Supposed to Have Arisen from Random Sampling." In this paper, he set the foundation for what is now called the chi-square test, a very popular significance test. However, Karl Pearson never extended his ideas to come up with a formal framework for testing in the way that Fisher did.

Karl's son, Egon Pearson, took the idea further. E. Pearson, in collaboration with Jerzy Neyman, came up with their own version of hypothesis testing. Whereas today the terms "significance testing" and "hypothesis testing" are used interchangeably, Fisher used only the term significance testing, whereas Neyman and Pearson used the latter. There were some differences in the two approaches, the most important one being the existence of two hypotheses in the method devised by Neyman and Pearson, which are called "null" and "alternative" hypotheses. Fisher battled fervently against Neyman and Pearson, but the version of hypothesis testing now used in practice is a blend between the two methods.

Hypotheses tests are everywhere in science. In fact, some disciplines are deemed scientific solely because the methods of statistics let us confirm or disprove hypotheses in a methodical and rational manner. Do you want to compare the average difference in a trait between two groups? Test the effects of an intervention in a school? Test the side-effects of a new pharmaceutical drug? Every time you read a study about the effects of food, substances, personality traits, or pretty much anything, there is a high chance there was a hypothesis test behind that. Basic statistics and testing are part of the curriculum for all young scientists.

So, we've established that hypothesis testing is very useful. However, as you might have gathered so far from this book, there is no free lunch. Hypothesis testing suffers from some serious problems that have caused a myriad of disputes, especially in the last few years. These issues touch upon some very important philosophical questions around the goal of significance testing and its use in science.

Fisher came up with the idea of statistical significance. He wanted to have some kind of way to declare whether a result was important. Therefore, he came up with a threshold, the famous threshold of 5 percent. If a p-value—which stands for "probability value"—of a test is less than 0.05, then the test is considered significant. If it is higher, then it is not significant. Some textbooks go further and even have different thresholds of significance. For example, a p-value less than 0.10 but higher than 0.05 is moderately significant. A p-value lower than 0.01 is extremely significant. These conventions have even passed into software packages for statistics. For example, the famous R programming language uses similar conventions.

Now, the curious reader will ask: *I see the value in using a threshold, but how did Fisher come up with 5 percent?* The truth is that this was an unsatisfyingly arbitrary decision that has come to haunt a large part of statistics and science for the last 100 years. In his book, *Statistical Methods for Research Workers*, Fisher writes:

> "The value for which P = .05, or 1 in 20 ... it is convenient to take this point as a limit in judging whether a deviation is to be considered significant or not. Deviations exceeding twice the standard deviation are thus formally regarded as significant."

This bound was later also used by Neyman and Pearson, thereby establishing it as a convenient standard.

Gosset himself was not a fan of the idea of significance. In a paper about this discussion, Ziliak[4] mentions that Gosset was more interested in the practical significance of the test, and not so much in whether the test itself was significant. Gosset had a more Bayesian approach to the whole debate, so he was focused more on decision making, rather than on uncovering some objective binary truth.

What the p?

The debate has heated up over the last few years. A book in 2008 by Ziliak and McCloskey,[5] called *The Cult of Statistical Significance: How the Standard Error Costs Us Jobs, Justice, and Lives,* outlines many of the problems with the current methodology. Ziliak and McCloskey give the example of the painkiller drug Vioxx by Merck. In clinical trials, the number of fatalities were five times higher than the comparison drug. However, because the number of observations did not reach the appropriate size, the results were not deemed statistically significant.

Another issue is p-value hacking. As early as 2005, John Ioannidis[6] discussed that most published researched findings are false. I'll let that sink in for a second. We've been building a case in this book around the powerful methods that humanity has created in order to control uncertainty. But now someone is telling us that a large part of the body of science is just plain *wrong*?

The British economist Charles Goodhart famously said, "When a measure becomes a target, it ceases to be a good measure." Most humans in their lives follow the principle of least resistance. They will try to find the easiest way to do things. This is why students cheat on exams, big consultancies abuse their reputation in order to close contracts, and creative accounting can make the financials of a country or an organization seem better than they really are.

The infamous p-values can make or break a career. Imagine you are a young researcher at a university—perhaps you actually are. You know that if you get a successful publication in a good journal, you could get the valuable scientific grant you need to carry out very important research. If you don't, then you'll end up without a

[4] Ziliak, Stephen T. "Retrospectives: Guinnessometrics: The Economic Foundation of" Student's" t." *Journal of Economic Perspectives,* 22.4 (2008): 199-216.

[5] Ziliak, S., & McCloskey, D. N. (2008). "The Cult of Statistical Significance: How the Standard Error Costs Us Jobs, Justice, and Lives." University of Michigan Press.

[6] journals.plos.org/plosmedicine/article?id=10.1371/journal.pmed.0020124

grant. You might have to find a job at a different university that is less prestigious. Besides prestige, this might mean that you have to move to a different city or country, something that you might not want to do. In the meantime, you might spend some months without earning an income. Therefore, there is a huge incentive to hack the p-value. This can mean anything from massaging the data, to running the same experiment multiple times until you get a p-value that is even marginally significant (e.g., 4.8 percent).

This has led to what is now called the *replication crisis*[7]—the failure of many scientists to replicate results published in other studies, which is easily explained by the fact that many of the results are actually the product of fiddling. A 2016 article in *Nature* on the topic[8] reported that 70 percent of the 1,500 scientists interviewed had failed to replicate at least one experiment conducted by some other scientist. In 2009, 2 percent of scientists admitted to falsifying studies themselves, and 14 percent admitted to knowing someone else who had falsified a study[9] (or perhaps they were answering for a friend).

A famous example of where p-hacking exists is parapsychology. While p-hacking in parapsychology might not be down to funding, but more due to a genuine desire to prove the existence of the paranormal, it nevertheless proves the point. Parapsychologists are studying phenomena like extra-sensory perception and telepathy. If you study the history of this subject, you will see two camps. The camp of parapsychologists argues that they have come up with some study for which they have found statistically significant results that prove X (e.g., telepathy). The other camp mentions that this is down to the experimenter running the same experiment 100 times until they get significant results.

This is a major issue with all sciences. When you say that the threshold of statistical significance is at 5 percent, then there is a 5 percent chance that you will get a false positive. A false positive occurs when you believe the result is significant, when really it is not. This is called a *Type I error* in hypothesis testing.

---

[7] The problem is so widespread that there is even an article about it on Wikipedia: `en.wikipedia.org/wiki/Replication_crisis`

[8] Monya Baker (2016). 1,500 scientists lift the lid on reproducibility, `www.nature.com/news/1-500-scientists-lift-the-lid-on-reproducibility-1.19970`

[9] Fanelli, Daniele (29 May 2009). "How Many Scientists Fabricate and Falsify Research? A Systematic Review and Meta-Analysis of Survey Data". *PLOS ONE*. 4 (5), `www.nature.com/articles/d41586-019-00857-9`

Daryl Bem, a distinguished professor of psychology at Cornell University, published[10] the results of seven experiments in the prestigious *Journal of Personality and Social Psychology*.[11] His experiments caused shock and much debate. Bem's experiments all concerned the existence of extra-sensory perception (ESP). In Experiment 1, the participants (consisting of 1,000 students at Cornell University) had to choose between two curtains. After they chose a curtain, a random mechanism decided, randomly, to display an image. This image was either an erotic image or some other non-erotic image.

The erotic images were considered to be positive reinforcement. The purpose of this experiment was to measure precognition, the ability to see events into the future. If precognition doesn't exist, then students would choose the curtain with the erotic image 50 percent of the time (since the choice of an image was completely random). If precognition really exists, then we would expect to see that the accuracy of predicting the erotic images would be more than 50 percent. It would be as though the students can look slightly into the future and predict where the erotic image would show up. The future position of the erotic images was selected with a 53.1 percent success rate.

In Experiment 2, Bem followed a different setup. The students were shown two images that were a mirror version of one another. Hence, there was no motivation to choose one image over another based on how they look. When a student chose an image, the computer would flash, for a few milliseconds, either a negative or a positive image. The choice of this image happened at random.

This experiment is based on the practice of subliminal messaging. If students had to choose between two images, but we flash positive images for a few milliseconds behind one image, and we flash negative images behind the other image, then our brain will tend to favor the image with the positive subliminal messaging.

So, what Bem did was to test whether causality can be reversed. We know that flashing images subliminally before, or while, an image is displayed can influence future decisions. What happens if we flash those images after the image is displayed?

In Experiment 2, Bem found that 51.7 percent of the students chose the images where the positive image was briefly flashed. Hence, according to his theory, it is as if students could briefly look into the future and receive positive or negative reinforcement.

---

[10] Bem, D. J. (2011). "Feeling the future: Experimental evidence for anomalous retroactive influences on cognition and affect." *Journal of Personality and Social Psychology, 100*(3), 407-425.
[11] www.apa.org/pubs/journals/psp/index

The rest of the experiments reported by Bem were quite similar, and they were deemed statistically significant. This provided proof, according to Bem, that ESP is real. Obviously, this caused a backlash. While there was no way to refute that the results were statistically significant, does this *really* prove that ESP exists? To his credit, Bem encouraged other researchers to replicate his experiments. None managed to replicate the research. Chris French wrote about his ordeal in *The Guardian*[12], and all the rejections the replicated study received, until it was finally published in *PLOS ONE*.[13]

What could have happened here? One possible issue is that the results by Bem were clearly false positives. Due to random chance, this is entirely plausible. However, seven false positives in a row are unlikely. So, something else that might have happened was that Bem might have had conducted multiple experiments and chose to selectively present only the positive results. There is always a genuine chance that ESP really exists.

However, extraordinary claims require extraordinary proof.

A Bayesian argument against hypothesis testing in this context was that Bem should have used a Bayesian method, which places a high prior that ESP doesn't exist. However, the degree of subjectivity involved in setting this prior probably raises more questions than provides answers.

So, what is the lesson learned? Let's assume for a second that ESP does indeed exist. What is the best method to go about proving it? Significance testing provides ways through which the results can be hacked. The only way to be sure that the results are correct is through replication. A single test based on a binary threshold won't do that. This is especially true for a much disputed and complicated topic such as ESP, whose existence would force us to revise the laws of physics.

This is further complicated by the possibility of small, but significant effects. Jessica Utts[14] has been a firm believer in the existence of remote viewing. Jessica Utts is a prominent statistician and also served as the president of the American Statistical Association in 2016. In 1995, she participated in an expert panel whose goal was to investigate the potential of remote viewing for espionage applications, based on the Stargate Research project that ran from 1978 until 1991. This panel was organized by the American Institutes for Research. The second person on the expert panel was Ray Hyman, professor emeritus of psychology at the University of Oregon.

---

[12] www.theguardian.com/science/2012/mar/15/precognition-studies-curse-failed-replications

[13] journals.plos.org/plosone/article?id=10.1371/journal.pone.0033423

[14] www.ics.uci.edu/~jutts/

According to Utts, there is consistent evidence that psychic powers exist, which can be confirmed by small, but consistent effects. This could be explained by psychic abilities following the normal distribution, much like IQ, with the majority of people displaying barely any potential and a few individuals displaying above-average abilities.

Is this theory correct? We don't know, and once again, this is a much-debated topic. However, the point is that the thresholds of statistical significance are *not* helping, since they are often used as "points" by researchers. Ten marginally statistically significant results are considered "proof," whereas ten results that are above the level of statistical significance are considered irrelevant. This doesn't help in the debate of complicated subjects, whose very existence could cause a paradigm shift in our world view.

The voices against p-values have become stronger over the years. In 2019, the most prestigious scientific journal, *Nature*, published a comment[15] against the use of thresholds of statistical significance. More than 800 statisticians signed it. It has become clear that there are social arguments against using it. However, the comment also raises some scientific concerns. Besides the issue of false positives, discussed earlier, another issue is false negatives. A false negative takes place when we refute the hypothesis, while it is true. This is called a *Type II error* in hypothesis testing.

Let's say that you are studying the effects of two drugs on blood pressure. One drug can reduce blood pressure anywhere from 1 to 10 points, with a mean of 4 points. Another drug can reduce it anywhere from 0 to 5, with a mean of 4 points, as well. As the test of the second drug will include some trials where blood pressure was not affected, it is very likely that it will be considered a non-significant result.

So, which approach should we follow? A better way than significance testing is to focus on decision making. Fisher's philosophy was based on a view that there is some kind of objective truth hidden behind the data—much like in Plato's idealism. While Fisher started off as a Bayesian, he ended up explicitly rejecting the Bayesian view of probability as being too subjective. Whereas Fisher never really expressed any official opinions on epistemological matters, all scientists, to one extent or the other, are influenced (either consciously or not) by a certain viewpoint in epistemology. Fisher's

---

[15] Amrhein, Greenland, McShane (2019). Scientists rise up against statistical significance, www.nature.com/articles/d41586-019-00857-9

viewpoint was probably a more romantic one, inspired by the hunt for objective truth. In that aspect, he wasn't much different from other great scientists, who also sought theories to explain the nature of the world.[16]

However, the criticism of significance testing and the gradual increase in popularity of Bayesian methods has given way to a more pragmatic use of statistics. In the modern day and age, data is everywhere. There are so many opportunities to use data, especially in a business setting. A large part of the application of data science and related research happens not in academia, but in a setting where everything is measured by profit.

When we view things under this spotlight, significance testing might seem fundamentally limited. Even if significance testing was flawless, it would tell us something about the world, but it wouldn't directly help us make decisions. There is a different way to use statistics to extract information from data, while at the same time offering greater flexibility and aiding with decision making. This can be achieved through the use of statistical models.

Statistical Modeling: A Useful Abstraction

"All models are wrong, but some are useful."

—George Box (1919–2013), one of the greatest
statistical minds of the 20th century

A model is simply an abstraction of the world. Nothing more, nothing less. All models are wrong, but some are useful, in the sense that they capture the right amount of abstraction between the real underlying phenomenon, our data, and our theories.

---

[16] This is a view that first appeared during the era of the Enlightenment and resurfaced in various forms over the years, such as positivism. Physics and chemistry were seen as the prime sciences leading this reductionist approach. However, this approach has been criticized over many accounts: From Heisenberg's uncertainty principle (which places a limit to our knowledge), to the inability of reductionism to explain emergent phenomena.

The principle of statistical modeling is really simple, even if the mathematical details can sometimes seem daunting and esoteric. We have two set of variables: the independent variables and the dependent variables. These are also called the covariates (or predictors) and the response variable.[17] The model sits in the middle and connects the covariates with the response, as shown in Figure 7-4.

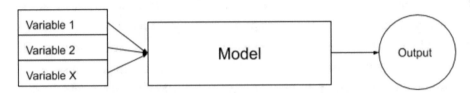

Figure 7-4. *Conceptual overview of a statistical model. In practice, most models follow this principle. You feed input into the model, and then you get an output, which consists of one or more responses*

Let's consider a business example: You believe that there might be some relationship between the demographic variables of your users and their purchase behavior. Let's say, for example, that you are an online retailer selling clothes. You believe that gender, age, and income might influence how much money customers spend on clothes. Using a statistical model, you can do the following:

1) Determine the expected spend for different types of users.

2) Understand which variables are significant.

The second point is particularly interesting. We can use significance testing within statistical modeling to assess the significance of individual variables. This makes statistical modeling very powerful. In fact, there are statistical models that can be seen as generalizations of significance tests. For example, the famous ANOVA test (as a reminder, ANOVA stands for analysis of variance) can also be done through linear regression—the most basic statistical model.

---

[17] In machine learning literature, exactly the same concept is described as input variables (or features) and output data (or target variable). In some cases, we can have more than one target variable, but the concept is always the same.

So, why don't we use statistical modeling all the time? There are a few reasons. First of all, simplicity is often key. Sometimes we are genuinely interested in a specific question that can be answered through significance testing. Hypothesis tests can be simpler to use than models. Statistics is used by many sciences, and many times fledgling scientists are trained only in the particular tests and models that are used most often in their discipline. If a t-test is all someone needs to prove their theory, then they won't care to go further than that.

Secondly, statistical modeling has a different focus than hypothesis testing. In science quite often we care about proving or disproving a hypothesis, and the way we prefer to do that is through experimentation. Fisher himself had contributed a lot to the theory of experimental design. A properly designed experiment lends itself nicely to hypothesis testing, and statistical modeling might be more than is required, once a hypothesis has been proven or disproven.

Finally, we didn't always have an abundance of data and computing power like we currently have. Fitting a model these days can be as easy as two lines of code. It certainly wasn't always like that. A few decades ago, we were employing statistical tables, aided with precomputed values related to the calculations required by hypothesis tests, which made the calculation easier. The smaller number of calculations for most tests, as compared to statistical models, was fairly convenient, which could have stirred many scientists toward the use of hypothesis testing.

Figure 7-5 shows an example of table of z-values from the standard normal distribution. The cells represent the proportion that lies between two z-values. So, the highlighted/underlined cell in the figure is the proportion of subjects that lie between 0.0 (the mean) and 0.04. These tables, for all kinds of distributions, were used in calculations for significance tests.

Z	0.00	0.01	0.02	0.03	0.04	0.05	0.06	0.07	0.08	0.09
0.0	0.0000	0.0040	0.0080	0.0120	0.0160	0.0199	0.0239	0.0279	0.0319	0.0359
0.1	0.0398	0.0438	0.0478	0.0517	0.0557	0.0596	0.0636	0.0675	0.0714	0.0753
0.2	0.0793	0.0832	0.0871	0.0910	0.0948	0.0987	0.1026	0.1064	0.1103	0.1141
0.3	0.1179	0.1217	0.1255	0.1293	0.1331	0.1368	0.1406	0.1443	0.1480	0.1517
0.4	0.1554	0.1591	0.1628	0.1664	0.1700	0.1736	0.1772	0.1808	0.1844	0.1879
0.5	0.1915	0.1950	0.1985	0.2019	0.2054	0.2088	0.2123	0.2157	0.2190	0.2224
0.6	0.2257	0.2291	0.2324	0.2357	0.2389	0.2422	0.2454	0.2486	0.2517	0.2549
0.7	0.2580	0.2611	0.2642	0.2673	0.2704	0.2734	0.2764	0.2794	0.2823	0.2852
0.8	0.2881	0.2910	0.2939	0.2967	0.2995	0.3023	0.3051	0.3078	0.3106	0.3133
0.9	0.3159	0.3186	0.3212	0.3238	0.3264	0.3289	0.3315	0.3340	0.3365	0.3389
1.0	0.3413	0.3438	0.3461	0.3485	0.3508	0.3531	0.3554	0.3577	0.3599	0.3621
1.1	0.3643	0.3665	0.3686	0.3708	0.3729	0.3749	0.3770	0.3790	0.3810	0.3830
1.2	0.3849	0.3869	0.3888	0.3907	0.3925	0.3944	0.3962	0.3980	0.3997	0.4015
1.3	0.4032	0.4049	0.4066	0.4082	0.4099	0.4115	0.4131	0.4147	0.4162	0.4177
1.4	0.4192	0.4207	0.4222	0.4236	0.4251	0.4265	0.4279	0.4292	0.4306	0.4319
1.5	0.4332	0.4345	0.4357	0.4370	0.4382	0.4394	0.4406	0.4418	0.4429	0.4441
1.6	0.4452	0.4463	0.4474	0.4484	0.4495	0.4505	0.4515	0.4525	0.4535	0.4545
1.7	0.4554	0.4564	0.4573	0.4582	0.4591	0.4599	0.4608	0.4616	0.4625	0.4633
1.8	0.4641	0.4649	0.4656	0.4664	0.4671	0.4678	0.4686	0.4693	0.4699	0.4706
1.9	0.4713	0.4719	0.4726	0.4732	0.4738	0.4744	0.4750	0.4756	0.4761	0.4767
2.0	0.4772	0.4778	0.4783	0.4788	0.4793	0.4798	0.4803	0.4808	0.4812	0.4817

Figure 7-5. *Example of table of z-values from the standard normal distribution (the standard normal distribution has a mean of 0 and standard deviation of 1)*

These days there are countless models for all sorts of problems and challenges. The key component of statistical models is that they are stochastic. That is, they explicitly model uncertainty. The way they do this, and the assumptions they make in order to achieve that, can vary widely from model to model, and of course there are very big differences among the different kinds of approaches.

The most popular, easy, and common model by far is linear regression. Linear regression has been the workhorse of science for a long time. A search for linear regression in Google scholar at the time of writing returns nearly five million results.

Linear regression follows a very simple line of thinking. Let's say that you have a scatterplot of data, as shown in Figure 7-6.

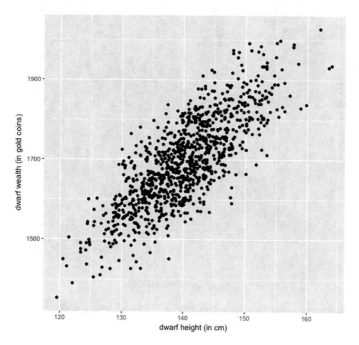

Figure 7-6. *Dwarf height vs dwarf wealth in Middle Earth*

Figure 7-6 shows the relationship between the height and the wealth of dwarves in Middle Earth. In this fictional example, there is a strong relationship between the height of a dwarf and their wealth, with taller dwarves seemingly being wealthier on average. It looks as if the relationship between the two variables follows a straight line.

Of course, there are other kinds of relationships that you might be able to identify, but a straight line seems entirely plausible in this case. Plus, a straight line is the simplest kind of geometrical relationship that exists, and this gives it a certain appeal. If you remember from earlier in the book, we discussed Occam's razor and our preference for simpler rules over more complicated ones. If we could draw a straight line to describe the relationship between the two variables, then we have a parsimonious and simple way to predict wealth from height.

There are multiple ways to draw a straight line. Just take a look at Figure 7-7. Which line is the best? It looks like the line in the middle captures the trend of the data a bit better than the lines that are on the edge. But is there a way to formulate this intuition?

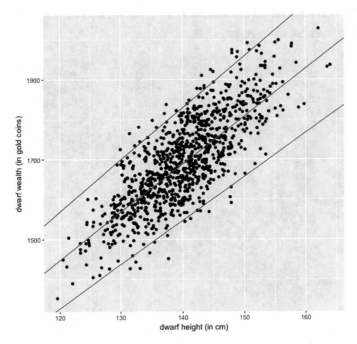

Figure 7-7. *Different ways to draw straight lines in a way that connects height with weight*

This intuition is captured through linear regression. The goal of linear regression is to come up with the line of best fit. That is, the line that most closely follows the trend of the data. So, you find the line of best fit, and it looks like the line shown in Figure 7-8.

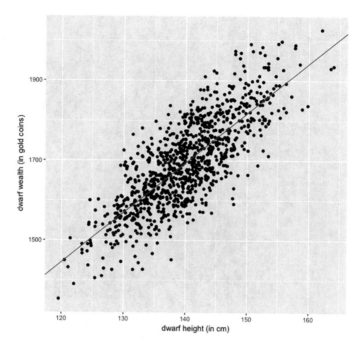

Figure 7-8. *The line of best fit for the height vs wealth problem*

The curious reader might have raised an eyebrow here. First of all, this line seems to follow the overall trend, but the relationship between height and wealth is not flawless. In fact, only a small percentage of the total points pass through the line. Secondly, what is the rule that we used in order to come up with this line?

As you noticed, there are some errors that the model makes. What can we do about that? We can't predict each individual error in the model. We accept that the model is going to make some errors. These errors are either due to epistemic or aleatoric uncertainty, but in this instance, we really don't care about the source. What we care about is the shape of the error. So, what does the error look like in this case? We can gather all the errors and plot them in a histogram. This is shown in Figure 7-9.

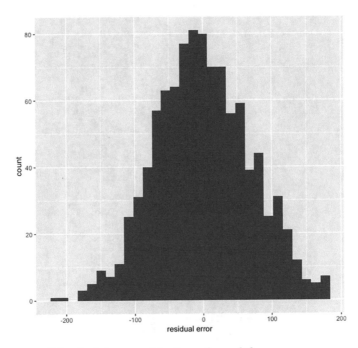

Figure 7-9. *Errors of the height-wealth dwarf model*

The errors start looking somewhat like the normal distribution. If you have noticed something from the book so far, we can model uncertainty using a probability distribution. Hence, if our errors look like a normal distribution, we can use this as a heuristic to guide us while we are finding the line of best fit.

In linear regression, we assume that our model will commit errors and that the errors are distributed according to a normal distribution. Middle Earth dwarves aside, why the normal distribution? Because the normal distribution is ubiquitous, and quite often the error of measurement is normally distributed.

Making these two assumptions (linear relationship and normally distributed error), we have the linear regression model. But what if we change these assumptions? Changing those assumptions will give us new models. John Nelder (1924–2010) and Robert Wedderburn (1947–1975) were the statisticians who came up with the concept of the generalized linear model. Both worked in the Rothamsted Experimental Station just north of London, the same place where Ronald Fisher worked when he developed many of his theories for experimental design.

Mathematicians like generalizations of existing concepts. For example, the rational numbers (e.g., numbers like 1.23) can be seen as an extension of the integers, since the set of real numbers also contains integer numbers. The set of real numbers can be seen as a further extension, because it includes both rational and irrational numbers (like $\sqrt{2}$). The history of mathematics is full of similar examples. The motivation to do this is that higher order relationships can help us uncover more insights and rules about how the world of mathematics works, and these can in turn be used in other models and theories.

Statistics is no exception to this rule. Nelder and Wedderburn managed to come up with a theoretical framework to merge some existing models (like logistic regression and Poisson regression) under a generalization of the linear regression concept. So, the generalized linear model assumes that the relationship between the predictors and the response is a linear one. However, the actual error can follow some other distribution. The famous Poisson model assumes, among other things, that the response variable consists of integers. When is this model relevant? Well, sports, for example. Scores in sports consist of integers. Hence, if we wanted to predict the number of goals in a football match, it would make more sense to use Poisson regression, rather than predict real numbers using linear regression.

Logistic regression can be used to model binary outcomes. Will someone buy (or not) a particular item? Will someone default or not on a loan? But what about the case where we have multiple potential outcomes? For example, in some sports, the outcome can be win/loss/draw. In this case, we can use multinomial logistic regression, which is a generalization of logistic regression to multiple outcomes. Beta regression can be used to model numbers in the range between 0 and 1, like proportions. Gamma regression can be used to predict positive real numbers. And these are only some of the models of the generalized linear model family.

Other kinds of models include survival models. Let's say that you are studying the reliability of a machine. If you buy a car now, how long until it needs its first repair? Or you might care about the survivability of patients suffering from a dangerous disease. What is the probability that someone will be alive in five years? Survival analysis helps find answers to these questions.

The names that have been associated the most with survival analysis are David Cox (1924–2002) and Waloddi Weibull (1887–1979). Cox gave us one of the most popular models in this area, called the proportional hazards model, whereas Weibull studied the distribution of the same name. The Weibull distribution is a good example of the kind of flexibility that statistical models can help express. Some entities have a higher probability of dying the more they have lived. This is true for most living creatures above a certain age.

However, for other entities, that's not the case. Businesses are less likely to go bankrupt the longer they've been around, as they are building a loyal user base. Hence, someone could argue that the longer a business has been around, the more likely it will survive even longer. Finally, there are some objects that have a constant probability of survival across their lifetime. For example, a glass vase that sits on the table will stay there forever, unless someone drops it. It doesn't matter whether the vase has been there for 1 year or 100 years.

Figure 7-10 shows the failure rate for different parameters of the Weibull distribution. The failure rate is the number of failures per unit of time. In this example, we are measuring the unit of time in a month. Hence the failure rate represents the number of units that are expected to fail each month. It is easy to see that the Weibull distribution is very flexible and can accommodate all of the scenarios discussed here.

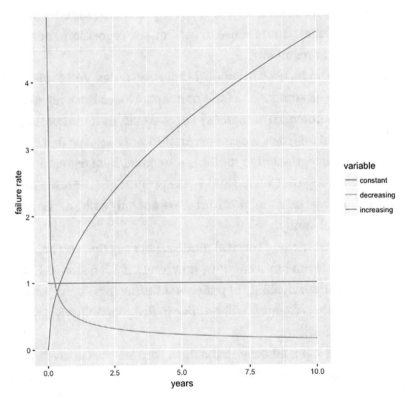

Figure 7-10. *Failure rate for different parameters of the Weibull distribution*

But what about non-linear relationships? For the most part, models in statistics have been additive and linear. The reason is that non-linearity is a Pandora's box. Once you start exploring it, you end up with myriad possibilities—and with them, myriad problems.

First of all, the interpretation of the models becomes more difficult. For example, let's revisit the dwarf linear regression model. Let's say that the linear regression equation is the following:

$$gold\ coins = 12 \times height$$

This model has an easy interpretation. For every centimeter in height, a dwarf gets 12 more gold coins on average, other things being equal. Straightforward and easy to understand, right? However, let's say that there was some other relationship that better explained the data and it was the following:

$$gold\ coins = 12 \times log\left(height\right)$$

How is this relationship interpreted? Now we have a logarithm, so the relationship is not as clear as it was before. For someone trained in a quantitative discipline, it's not difficult to understand, but for the average person it might be more difficult to visualize in their heads what a unit increase in the logarithm of height looks like.

Let's say that we want to add a new variable—beard length—into the equation. And we come up with this:

$$gold\ coins = 12 \times \frac{log\ log\left(height\right)}{\sqrt{beard}}$$

This is even more difficult to understand. But what if the equation was like this one:

$$gold\ coins = 12 \times height + 11.5 \times beard$$

This model can be interpreted as follows. For every centimeter increase in height, given no change in the length of the beard, a dwarf has 12 more coins on average. For every centimeter increase in the length of the beard, given height doesn't change, a dwarf has more 11.5 gold coins on average. Much easier to understand, right?

Secondly, there are many different non-linear functions someone could use. You could easily end up with two or three different models, with difficult interpretations, but similar performance. Statistics always had a strong focus on interpretability and transparency of the results, and non-linear models lose on that front.

That being said, there are non-linear models that are used in practice. The generalized additive model is one example. However, the domain of non-linearity is mostly dominated by machine learning methods, about which you read more later.

Statistical models and tests are very powerful tools, and they have served us well for a long time—some for centuries. However, as mentioned earlier, when the only tool you have is a hammer, everything starts looking like a nail. Statistical models have often been misused, partly because of the fact that in many sciences, the fledgling scientists do not have enough time to study statistics in depth. Therefore, they often miss out on important assumptions behind the models or are oblivious to alternatives. So, what's wrong with statistical modeling?

There are two common fallacies that I have observed over the years. The first one is the Gaussian fallacy. The second one is the linear world fallacy.

The Gaussian fallacy refers to the overuse of linear regression as a model for anything. Linear regression is the first statistical model that anyone learns and it is included in all statistical packages. However, the world is not always Gaussian. There are a multitude of other models that can cover the non-Gaussian case, like Poisson regression, logistic regression, generalized additive models, and many more, but these are often ignored.

The linear fallacy assumes that everything in the world is linear. For better or for worse, it is not. The vast majority of statistical models employed assume the relationship between input and output is linear. In Figure 7-11, you can see an example of a non-linear relationship between two variables. Let's look at the relationship between the span of the wings of dragons and their speed.

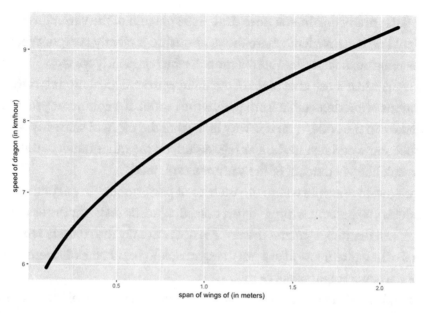

Figure 7-11. *Example of a non-linear relationship between a set of two points*

Say that we want to run linear regression to build a model that predicts the speed of a dragon from their wingspan. The regression model explains around 94 percent of the variability. This sounds like a good model. So, let's plot the predicted values (red line in Figure 7-12).

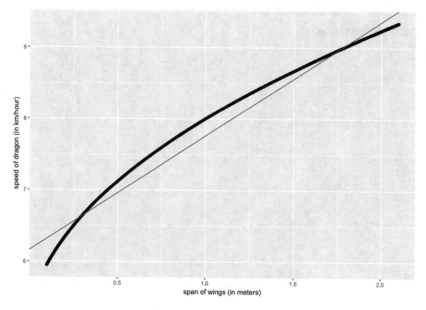

Figure 7-12. *Regression model on the same data*

The model is pretty horrible. It does explain 94 percent of the variability, but it tries to fit a straight line in a problem where the relationship is clearly curved. What we have done here is inflate expectations about a model, which is clearly wrong.

Why do issues like these take place? One issue with statistical models is that violating the assumptions is not detectable from within the model. If a computer programmer makes a mistake in the code, then it is very likely that the piece of software won't run, or it will produce some odd results. Linear regression, on the other hand, will still run and produce a result like "48 percent of the variance explained."

Here you saw an example where the metrics of performance are actually quite good, but the model is 100 percent wrong. In this case, diagnostic tests, for the assumptions of the model, will demonstrate that the error is not normally distributed. Hence, the assumptions of linear regression are not being met, and we need a different model. But not every practitioner might observe this.

This example might seem abstract, until you realize that many of these issues are used in models that can affect our lives. Pharmaceutical drugs are approved or denied based on these models. Econometric models provide forecasts for the economy, so that central banks can adjust inflation rates. Applying all these models in the right way is challenging.

What if there were methods that were easier to use, that didn't have all these assumptions, and that could deal with non-linearities in an easier way? Wouldn't this make modeling and predicting easier?

Enter machine learning.

CHAPTER 8

Machine Learning: Inside the Black Box

"I believe that at the end of the century the use of words and general educated opinion will have altered so much that one will be able to speak of 'machines thinking' without expecting to be contradicted."

—Alan Turing

Alan Turing was not wrong.

Just a few years later, in 1959, the well-known American computer scientist Arthur Samuel came up with the following, widely accepted definition: "Machine learning is about giving computers the ability to learn without being explicitly programmed."

No other technology right now is getting more hype than machine learning, and rightfully so. In 1991, the World Wide Web became publicly available, connecting everyone on the planet. Amazon launched in 1994. Blogs were invented in 1997. Google launched in 1998. Facebook was founded in 2004. Amazon released its cloud computing services in 2006. In 2007, the first iPhone was released.

Since the 90s, we've seen a constant trend of interconnectivity, data generation, and data storing. We express our thoughts online, we shop online, we ask questions online about anything (from directions to where to go out to eat), and all this data is constantly being recorded, building a digital representation of our world. This has become even more prominent after the COVID-19 pandemic, during which, many professionals were made to work from home.

In this data deluge, there is one particular set of tools that started rising in popularity in the 90s, and that has now dominated the world of technology: machine learning. Machine learning is now everywhere. From smartphones, to online shopping, to public

© Stylianos Kampakis 2023
S. Kampakis, *Predicting the Unknown*, https://doi.org/10.1007/978-1-4842-9505-2_8

transportation—machine learning algorithms are constantly measuring, predicting, and optimizing responses. Compiling a complete list of use cases for machine learning is not only beyond the scope of this book, it would probably require a book on its own.

What started as an effort to re-create intelligence has become the most powerful tool we currently have against uncertainty of all kinds. No treatment of uncertainty would be complete without a conversation about machine learning. The purpose of this chapter is to demystify machine learning and some other terms like "data science." The chapter looks at its history and explores why machine learning seems to work so well—but has several shortcomings, too.

Data Science and History of Machine Learning

These days, the trendiest word is not "machine learning" or "statistics," but *data science*. This section explores this term further before moving on to talk about machine learning.

What is data science? Data science is essentially a fancy new name devised by the world of industry to merge various disciplines related to data analysis: machine learning, statistics, computational intelligence, data mining, artificial intelligence, and more.

Since the beginning of time—or thereabouts—humans have tried to re-create intelligence inside machines, to analyze data and to control uncertainty (among other things). This has given rise to various scientific fields over the decades, each with a different methodology as to how to achieve these goals. Some of those fields came up with radically different approaches, but many others came up with very similar methods.

For example, computational intelligence was very popular in the 1990s (but not so much anymore) and was based on a biological analogy. The theory was that if nature has developed mechanisms that display intelligence, then we might be able to replicate intelligence in a machine by mimicking those mechanisms.

Are you convinced?

It makes sense to a certain extent. If ants can organize themselves inside a colony, maybe we can learn something from them. If our brains consist of neurons, then we can mimic their functioning and use it to re-create intelligence.

However, this approach didn't get too far. The biological analogy seemed to be too loose, especially for mathematically oriented scientists. Computational intelligence fell out of fashion, and machine learning started taking over.

Classic artificial intelligence was associated with an approach developed in the 50s that focused on using logic rules and reasoning in order to re-create intelligence.

This approach was associated with people like Allen Newell (1927-1992), Herbert Simon (1916-2001), John McCarthy (1956-), Marvin Minsky (1927-2016), and Arthur Samuel (1901-2000). Their approach considered logic to be the fundamental process of intelligence and believed that it would be possible to create an intelligent machine—as long as we could create a machine that could think in logical rules. This approach was also not very successful, and it was later abandoned.

Cybernetics is another term that had appeared earlier in the 40s, used by Robert Wiener to describe systems that could respond to their environment. *Data mining* and *pattern recognition* were also popular terms being bandied about in the 90s.

For most of history, these approaches and algorithms were being developed in academia, and in some tech companies, like IBM. However, after the data deluge toward the beginning of the 21st century, data-crunching algorithms suddenly started becoming more and more valuable. The fragmentation of those algorithms in different disciplines and schools of thought didn't help very much in selling products or services. Hence, the term "data science" was born, in order to encapsulate any data-related algorithm, method, product, or service.

This term has come to dominate the industrial landscape, largely because it is easier to say that you are a data scientist than to explain in depth the differences between the many schools of thought and which one you are aligned. I also introduce myself as a data scientist. People outside of the world of tech ask what is data science, and what does a data scientist do? And then they meet all sorts of answers. The truth is, there is a long historical tradition in all disciplines related to data analysis, uncertainty, and artificial intelligence. These traditions have been primarily based in some of the leading universities in the world, with some of the big technology companies also playing a role. Nowadays these firms are starting to surpass the universities, simply because they hold huge resources in their possession.

The term *data science* is useful for selling products and services, but when someone wants to understand the subject in depth, we have no choice but to actually come to grips with the term and study each subject separately.

If I had to define data science, I would say that these days, most data scientists mainly deal with the Python or R programming language. They most likely use techniques from both machine learning and statistics—a kind of cherry-picked combination, depending on their goals—in order to analyze and extract insights from data. However, as you'll learn in this book, machine learning and statistics have different historical origins—and different ways of approaching things.

Going back to the history of machine learning, it shares great parity with the history of classic AI. The approach of following rules of logic in order to re-create intelligence is called *the top-down approach.* This way of thinking goes like this:

1) Intelligence in a domain consists of a set of finite logical rules. For example, in the medical domain, a doctor has multiple rules of the form "if symptom X, then disease Y."

2) Merging many rules together should be enough to produce intelligence.

3) Hence, the objective of the AI engineer should be to encode these rules in the machine.

While this approach had some successes—like the MYCIN expert system developed at Stanford in the early 70s, which, at its time, was said to be more effective than junior doctors—it never went quite far enough. As with many cases in computer science, the people involved in this community overpromised and under-delivered; it's an easy trap to fall into. AI went through two periods called "AI winters" (a term first used in 1984), which were traditional tech bubbles. The size of the market at some point was worth billions of dollars, but then it became impossible to raise investment.

In the late 80s and early 90s, a different approach to creating intelligence grew in popularity, seemingly based on the exact opposite intuition. Instead of handcrafting rules for the machine to follow, let the machine learn on its own from data. This is a *bottom-up approach,* compared to the top-down approach adopted in classic AI.

The term "machine learning" was useful for two reasons. First, it accurately described what these algorithms were actually doing. Second, it helped de-associate this new approach with classic AI, which had grown toxic. The route taken in machine learning is similar to other approaches in the 90s, like computational intelligence, data mining, and pattern recognition. In fact, sometimes those terms could be used interchangeably.

So, come the 21st century, and the bottom-up, data-driven approach is dominating many parts of our lives. Why (and how) has it been so successful? There are two main reasons.

First, the top-down approach requires a lot of human input. The person who is crafting the rules has to think of every possible rule that could describe or affect a system. Perhaps needless to say, this is a very difficult, laborious, and time-consuming task. On the other hand, the bottom-up approach is much more efficient. The algorithm can learn everything from the data. The only thing that the human operator has to do is find the best parameter settings for the algorithm.

Secondly, there is a fundamental reason that the data-driven approach produces better results. In case you haven't realized it yet, our world is fundamentally uncertain. It is true that there are many cases where there is no uncertainty at all around an outcome or a type of relationship. However, in the majority of cases, there is. Let me give you an example. Try to come up with a set of rules to predict the price of a stock in the next two days. Maybe you can come up with a set of rules that work *approximately* well. But do you believe that anyone can come up with a finite set of rules that work well—every time without fail—down to decimal point precision level? Honestly, probably not.

Instead of pretending that uncertainty doesn't exist or trying to completely remove it from the equation, the best method is to simply accept it. Bottom-up approaches start learning from the outset that predictions will never be perfect, nor will they have to be. The objective is to simply reduce the number of mistakes.

This intuition is the reason that we now have the best version of intelligent algorithms, better than we've ever known before.

Choose Your Learning Type: Supervised, Unsupervised, Reinforcement, or Other

Machine learning is not a single, homogeneous family of algorithms that all do the same thing. Rather, as in life, there are nuances—there are different types of machine learning applications. The most popular ones are supervised learning, unsupervised learning, and reinforcement learning.

Supervised learning is the workhorse machine learning. It is by far the most popular category of algorithms out there. In supervised learning, the algorithm is faced with the following problem: there is an input and an output. You have to find a way to match the input to the output. The algorithm sits in the middle between the input and the output—like a black box—and determines how the two are connected. You can use supervised learning for many different things: from automating human tasks to predicting the future.

Think of spam mail detection, for example. Every time you report an email as spam, there is an algorithm that learns what is spam and what is not, automating your decision. Think of predicting the price of a stock. An algorithm can take into account multiple sources of information, such as historical trends and conversations on social media about different companies, and then use that to predict where the price will move to next. Algorithms track what products you purchase on online retailers, in order to find what you are most likely to buy next. When you upload a photo online, algorithms can automatically detect where the human faces are and ask you to tag them. All these are examples of supervised learning.

When most people think of machine learning applications, they are usually referring to supervised learning. Since the 90s, the field saw an explosion of powerful algorithms, which, in combination with the data deluge experienced around the turn of the millennium, turned machine learning into a game changer. The modern version of the support vector machine came out in 1992. Gradient boosting was discovered in 1999.[1] Leo Breiman (1928–2005) came up with the "random forest algorithm" in 2001.[2] Scikit-learn, the most popular machine learning library for the Python programming language, was created in 2007, turning Python into the standard language for machine learning. These days, there is even a competitor website, called Kaggle,[3] where people can compete against each other in supervised learning.

Supervised learning is closely related to statistical modeling. From the perspective of machine learning, statistical models are supervised learning. Actually, supervised learning employs some algorithms borrowed from statistics like linear regression and logistic regression.

The main difference between the two disciplines is in the *way* of thinking. Statistics were created in mathematics departments, where there is a strong focus on methodological soundness and transparency, so an abstract approach. On the other hand, machine learning was born in computer science departments and labs, where many people are engineers at heart and prefer experimentation and pragmatism over theory. Machine learning was born through tinkering and experimentation with a focus on predictive power, so an empirical approach.

In machine learning, the proof that a model is good resides in its performance on what is called a "test set." If the model has good predictive capability, then it is a good model. In statistics, it is not enough to check performance. You also need to test a series of assumptions. Machine learning has been attacked for its lack of theoretical

---

[1] Friedman, J. H. (February 1999). "Greedy Function Approximation: A Gradient Boosting Machine" (PDF).

[2] You can find the original paper online here: `www.stat.berkeley.edu/~breiman/randomforest2001.pdf`

[3] `www.kaggle.com/`

rigor. Even though there are some theories about how machine learning works (many of them inspired by Bayesian statistics), the truth is that machine learning has been more focused on performance and applications than theory—which many argue is a good thing. As the American engineer and esteemed management consultant William Edwards Deming (1900–1993) once said, "In theory everything is possible; however, I live in practice and the road to theory has been washed out."

Unsurprisingly, unsupervised learning is the exact opposite of supervised learning. In *unsupervised learning,* the algorithm is given a set of inputs, but there is no outcome to match these inputs to. Without output, there can be no supervision. Instead, the algorithm discovers patterns in the data. Let me give you an example. Say that someone showed you images of dogs, cats, and fish and asked you to categorize them. What categories would you come up with? Most people might just come up with three categories—cats, dogs, and fish. However, someone might argue that there are only two categories: furry mammals and fish. Or, maybe there are two categories, of different types to the previous ones: animals that are okay with water (dogs and fish) and animals that dislike water (cats).

Which one of those categorizations is the best one? The answer is that there is no objective criterion to judge by. Unsupervised learning is fundamentally an ill-defined problem. It can help you discover patterns in the data, but there is no objective way to declare which solution is the best one. Probably the most famous example of unsupervised learning is customer segmentation, which allows companies like retailers to group their customers into different profiles and then use this information for targeted marketing. I'm sure you can think of a time when you've clearly been categorized as a particular demographic, only to have advertisements directed at you that are way off the mark, just because you once bought your Great Aunt a garden gnome.

The French computer scientist Yann LeCun (1960–), one of the prime researchers in deep learning (as well as recipient of the Turing Award), has voiced the opinion that unsupervised learning is, possibly, the most important type of them all. The reason, he argues, is that for humans, unsupervised learning makes up the bulk of our experience. We learn most concepts not with direct supervision and reinforcement, but rather by observing patterns in the environment. Cracking the problem of generating true artificial intelligence—in that it is akin to human intelligence—might require us to crack the problem of unsupervised learning.

Reinforcement learning also lies at the more anthropomorphic end of the machine learning spectrum, and it closely corresponds to the human condition. In reinforcement learning, the algorithm is referred to as an "agent." The agent lives in an environment, which can be physical (as is the case with robots) or virtual (as would be the case with an algorithmic trading system). The agent is trying to maximize some kind of reward, while avoiding punishment. In order to do that, it can take a limited number of actions, it has access to the history of all past interactions with the environment, and it knows what the current state of the environment is like now. The agent learns by experimenting in the environment and by making projections of what the future will look like. Quite often, the agent will have to learn how to live with some punishment in order to get a greater reward later on.

How is this like real life? Well, think about going to college, for example. Going to college costs money and effort. Also, you have to consider whether the subject you are going to study is actually going to help you land a job. If you undertake this effort, it is likely that in a few years, you will find yourself in a much better position financially than what you would have been otherwise. From the perspective of reinforcement learning, you are trying to maximize financial gain in the future by experimenting and incurring some penalties in the short run.[4]

Or think about learning to ride a bike. You have to learn how to put your legs in the right place, and then balance your body and find the right pace. While you're doing this, you are very likely to make mistakes; you'll hurt yourself. But as time passes, the mistakes are fewer and fewer, and hurt less and less, until you become a decent rider. This is another example of reinforcement learning.

Reinforcement learning stole the spotlight in 2015. An algorithm called AlphaGo became the first computer program to beat a master professional in the game of Go. Believed to have been invented in China over 2,500 years ago, the game of Go is like a much more abstract and complex version of chess. Until the fateful match with Fan Hui in London in 2015, it was believed to be unbeatable by machines. AlphaGo was based on reinforcement learning.

There are many other subtypes of machine learning, like active learning and semi-supervised learning, and recommender systems. However, covering each one of these is outside the scope of this book. Also, new algorithms are being developed all the time, some of which might combine all these different types of learning. For example,

---

[4] Obviously, there might be other benefits in going to college, such as learning new things, living in a new city, making friends, or finding the love of your life. But I try to keep the example simple.

reinforcement learning can amalgamate supervised learning, and supervised learning can use unsupervised learning. Just to briefly mention one of the most exciting subtypes of learning—*one-shot learning,* where the objective is for the algorithm to learn with only one or a few examples. This is similar to the way humans learn some new concepts, such as faces.

All the methods outlined here are built for different uncertainty-based applications. In supervised learning, we assume that there is a relationship matching the input with the output, but what we don't know what this relationship looks like. In unsupervised learning, we assume that there is a pattern, some kind of regularity, but we don't know what this regularity is. In reinforcement learning, the agent faces uncertainty about the environment and the rules that govern it. Together, all these techniques provide a great toolbox to deal with a variety of situations where we face uncertainty.

The Bias-Variance Trade-Off

Another key concept in machine learning is the bias-variance trade-off. Let me explain.

Algorithms always face this particular trade-off, but it's not *just* algorithms. As you'll see later in this chapter, humans also face bias-variance—and potentially for the same reason as algorithms.

But let's first look at what bias and variance mean in the context of algorithms. By bias, I am referring to assumptions that the algorithm makes about the data or the problem. Heavily biased algorithms have a very specific idea of what the world looks like. In real life, heavily biased people will (likely) never change their opinion. For example, let's say you believe that all elves are smarter than dwarves. It would take myriad counterexamples for you to change your opinion. Algorithms are very similar.

Linear regression is an example of a heavily biased algorithm. It assumes all relationships in the world are linear. Imagine that someone shows you the image in Figure 8-1. They give you a ruler, and they ask you to draw a single straight line that follows the pattern of the data as closely as possible. Being forced to use a ruler, you are forced to use a straight line, and you will come up with something like the solid line in Figure 8-1.

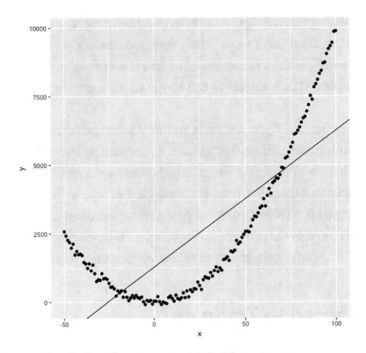

Figure 8-1. *Example of what happens in high-bias situations*

Low-bias algorithms are the opposite. In everyday life, you find people who are more open to new suggestions and ideas. These people have low bias. Algorithms are similar. Low-bias algorithms are open to letting the data change their opinion. For example, let's say someone posed the same problem to you as previously, but told you that you are not forced to use a ruler. You can freely draw something. You are not the best at drawing lines, but this gives you more freedom than previously. You come up with something like the line shown in Figure 8-2.

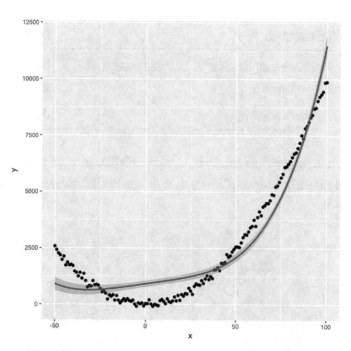

Figure 8-2. *Example of a low-bias algorithm*

This is much more accurate than the previous shape, but still not completely precise. In order to explain the full picture around learning, we also need to consider the variance.

Variance describes how much an algorithm is willing to listen to the data. Coming back to the real-life example, you might have encountered some people who change their minds all too easily. Right now, they believe that jazz music is the best there is. Next month they're into heavy metal. Or maybe they pick up a new hobby every other month. Or they just change their opinions about political subjects all the time, based on what they see in the media; they're a swayable, malleable audience. These are examples of high-variance people. On the other hand, people with low variance will not shift easily from their original position.

Bias and variance are linked. People with high bias and low variance will never move from their original opinions. However, people with high bias and high variance might start with a very strong opinion, but they are likely to change their view as more data comes along. People with low bias and high variance are just open to any old suggestion, to an extreme extent. These might be the people, for example, that are most susceptible to peer pressure. The following table summarizes all of this:

	High Variance	**Low Variance**
High bias	Will change their minds after they see enough counterexamples.	Never change their minds.
Low bias	Will believe anything.	Just the right balance.

As you've learned, algorithms often behave much like humans. In this context, the low-bias, low-variance algorithm will be just right and will produce a result like the one shown in Figure 8-3.

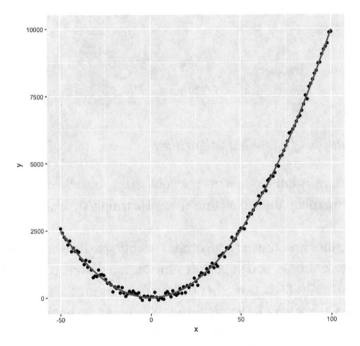

Figure 8-3. *Example of an algorithm that is just right*

High variance makes algorithms (and humans) confuse the noise with the signal. Not every news story you hear is true, and not every data point is free of noise. That is perhaps more widely spread now than it has ever been. Extracting information from the right sources is a balancing act, and, again like humans, machine learning algorithms have to tread this balance carefully.

Humans also use cognitive biases. Those biases have existed partly because they have served us well in a variety of situations in our evolutionary history. But, quite often, they can lead us to bad decisions. For example, when we still lived in tribes, being able to quickly detect whether someone was outside of our tribe was a useful skill to have. It was also normal to be skeptical of outsiders, given that our world was dominated by a small tribe, and any people outside of it would probably not benefit us much, and could even be hostile. But in a globalized world, being skeptical of other people for superficial reasons doesn't really assist us in anything. Whether we're talking about humans or machines, some principles of learning are the same.

Speaking in terms of uncertainty, high bias and low variance are fine when there is no uncertainty. Having a high-bias and low-variance opinion about the fact that the sun will rise tomorrow is probably a good thing. High variance and low bias are much better suited in those situations where uncertainty dominates. Having high variance and low bias that the growth rate of the economy will be 2 percent next year is a good thing. The economy might do much better than that—or it might just crash. Likewise, dwarves might outsmart elves (well, every now and then).

One of the ways in which machine learning algorithms improve their performance is through a technique called *regularization*. As you might have noticed in Figure 8-3, the "best" fit still contains some errors. It doesn't pass through every single data point, so it incurs mistakes, but overall it looks pretty good. However, the model in Figure 8-4 is perfect: there is zero error.

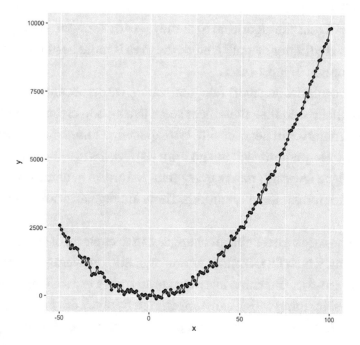

Figure 8-4. *Perfect model, as measured by error*

Would you prefer this model or the previous one? The last image is a prime example of a model that has very high variance. The model believes that every single datapoint with which it is presented is correct. It is most likely also wrong. In machine learning, we say that such a model "fails to generalize." This means that the model has memorized every data point, but it hasn't really *learned* anything. Let's look at another real-world example.

Let's say that you are learning algebra. Someone presents you with the following exercises:

$$2+x=0$$

$$5+x=1$$

The solution to the first exercise is -2, and the solution to the second one is -4. Following this reasoning, you might answer the following:

$$7+x=2$$

The solution is -5. But someone who has only memorized *solutions* will not be able to compute an answer and will respond with -2 or -4, because they have never seen the last formula before. It's a classic case of "computer says 'no.'" In machine learning, we say that they have failed to "generalize."

Regularization forces the algorithm to come up with a smooth solution, committing errors but generalizing better overall. A smooth curve is more aesthetically pleasing, and it is also the best option in most cases.

So, why does regularization work? Regularization works because our world is, overall, fundamentally smooth—that might seem like an odd claim for a data scientist in a book on uncertainty, but hear me out. We expect that tomorrow will look a lot like yesterday. The temperature will not go from -20 Celsius to +40 degrees overnight. Children grow taller a few centimeters every year. Nature, in general, does not like abrupt changes. It is fundamentally smooth. There are regularity and patterns in how things work.

There are some cases where abrupt changes can take place, of course, like in the case of complex systems, which I'll talk more about later. But, on average, the smoothness assumption has served the machine learning community very well, in a wide variety of problems. This is something which can, unfortunately, break down in the real world, when we face severe external shocks (like a pandemic or a financial crisis). But it has served us well for the most part.

Machine Learning vs Statistics: Why the "Dumb" Approach Works

"Simple models and a lot of data trump more elaborate models based on less data."

—Peter Norvig, Director of Research at Google

What's notable about machine learning is that it works so well with relatively little theory backing it up, contrary to statistics, which had been the dominant paradigm for data analysis. While there are some theories as to how and why machine learning algorithms work, the truth is that for a large body of machine learning research, the main driver has been tinkering and intuition, not theory—just as we saw earlier.

Let's say that you're faced with a problem of matching an input dataset to an output. For example, you want to create a model to predict an elf's magical ability from the magical ability of the elf's parents. The statistician would use a model with carefully crafted assumptions. Then, the statistician would proceed to test those assumptions and come up with tests and results that can tell you how well the model fits the data.

A machine learning theorist will use a model without any assumptions. The model will be based on data and tested on data. The model's ability to perform well will be measured only by the model's ability to correctly predict new cases.

Statistics is more theory and assumption-driven, and machine learning is—unsurprisingly—data-driven. In fact, many concepts in machine learning have been outright stolen from statistics, without much explanation of justification. For example, artificial neural networks (often abbreviated to ANN) used logistic functions as nodes for a long time, for the sole reason that these functions were already a popular choice in statistics. Decades after those functions were being used, it was discovered that they are better choices for neural networks.

The curious thing about machine learning is that through tinkering, the machine learning community managed to get unparalleled performance in a variety of tasks, unlike anything that had been witnessed before. Probably the prime example of this approach has been deep learning. Neural networks have been around as a concept for a long time. Actually, the first time someone thought of neural networks was when neurophysiologist Warren McCulloch (1898–1969) and mathematician Walter Pitts (1923–1969) developed a simple electrical circuit in 1943. The first modern neural network was the *perceptron*, devised by Frank Rosenblatt in 1958. Later advancements (mainly the invention of backpropagation) pushed neural networks even further. However, they fell out of fashion during the 90s, especially as other algorithms, which were better founded theoretically (like support vector machines[5]), showed up.

Neural networks experienced a renaissance due to the efforts of Geoffrey Hinton, Yann LeCun, and Yoshua Bengio, who came up with the concept of deep learning. The term "deep learning" is actually due to Hinton, who came up with it as a ruse to trick academics into publishing his work—papers on neural networks had fallen so out of fashion, they weren't even being published. If ever there was a story that illustrates the irrationality that the academic establishment often displays, this is it!

In any case, deep learning solved problems that had not been answerable in the past. And most of these were exceptionally difficult problems—they related to computer vision, audio, language, and so on. Deep learning has dominated the landscape of any complicated problem for which we have large amounts of data. All the big technology companies are investing huge amounts of money in furthering deep learning research. Hinton, LeCun, and Bengio correspondingly became superstars and were awarded the Turing Award[6] for their contribution in computer science.

The interesting thing about neural networks is that they are based on a loose biological analogy—you might have guessed this from the name. The core concept behind neural networks was that by imitating the way that biology creates intelligence, we could re-create intelligence inside a computer. Neural networks didn't have much theory backing them up, and this is one of the reasons that they fell out of fashion in the 90s, when algorithms based on more sound theories were the trend. Ironically enough, one of the few theoretical papers about neural networks had concluded that in neural networks, you don't need more than one hidden layer.[7] This was called the "universal approximation theorem."

The core innovation that deep learning introduced was that by adding more layers, the networks became much more powerful, in contrast to what the universal approximation theorem had predicted. So, what was discovered by tinkering actually ended up being far superior to what the theory had initially allowed. Now, there are

---

[5] Support vector machines were created by Vladimir Vapnik, whom I talk about later in the book.
[6] Named after Alan Turing, the Turing award is like the Nobel prize of computer science.
[7] Cybenko, G. (1989) "Approximations by superpositions of sigmoidal functions", *Mathematics of Control, Signals, and Systems,* 2(4), 303–314.

many mathematicians and theoreticians trying to come up with explanations as to why deep learning works so well. It's likely that some of these explanations will have their own shortcomings, and then some other tinkerer will disprove the theories wrong.

While the discipline of statistics still shines as the best solution in many applications, machine learning, and especially deep learning, is the *only* available solution for a wide variety of complex problems. The reason might be what Nassim Nicholas Taleb has called the "lecturing birds on how to fly"[8] effect. Most scientific textbooks present science as a rational and organized endeavor. However, they are full of after-the-fact explanations. Scientific progress, Taleb argues, requires a combination of skill and luck, and it is more often the result of tinkering, experimentation, and happy accidents, rather than careful planning. What might have happened if Newton wasn't sitting under that apple tree? Machine learning has treated the problem of uncertainty as a problem to be solved first through experimentation, rather than theory. In doing so, it has worked amazingly well on a variety of subjects. The "lecturing birds on how to fly effect" in uncertainty exists in cases where the practitioner expects the world to follow preconceived models. This approach quite often fails. This is also why Amazon advocates "bias for action."[9] In their own words: "Speed matters in business. Many decisions and actions are reversible and do not need extensive study. We value calculated risk taking."

But does this mean that we need to throw theory out of the window? Not at all. Both approaches have their merits. The point I am trying to demonstrate here is how research against uncertainty itself faces uncertainty. It is a truth universally acknowledged. Uncertainty cannot be avoided, but it can be dealt with. We just need the right approach. Hence, endless theorizing can often be counterproductive, as in the face of severe circumstances, not taking action can often be disastrous.

Machine Learning Shortcomings

Is machine learning perfect? Well no, even if many service providers try to convince you that it is. Machine learning has been heralded as a solution to everything, but it has some serious shortcomings.

---

[8] www.edge.org/conversation/nassim_nicholas_taleb-understanding-is-a-poor-substitute-for-convexity-antifragility
[9] www.amazon.jobs/en/principles

The first shortcoming is that machine learning is not integrated with logical reasoning. While the classic AI approach didn't go far enough, logical reasoning, nonetheless, is part of human reasoning. If I provide you with these statements:

1) All elves can throw lightning bolts.

2) Bob is an elf.

Then you can conclude, rightfully so, that you shouldn't mess with Bob, because he can throw lightning bolts. However, a machine learning system would have to see multiple examples of similar statements before it can understand the relationships. There are research efforts currently underway tasked with answering using neural networks, where the objective is for a network to learn how to answer questions and read logical statements. However, the amount of data required for neural networks to perform these tasks does not make them look too intelligent, as this ability comes naturally to us. It's like the example of a computer who can identify a cat video—which sounds incredible until you remember that two-year-old children can identify cats.

Another issue with machine learning is that it is opaque. Algorithms are famously treated as black boxes. It is difficult to understand what happens *inside* an algorithm, and how it makes decisions. That being said, there are ways to extract insights from the inner workings of an algorithm. Some algorithms will let you know which variables they consider to be the most important. There are methods, like Shapley value explanations, that can help users understand the interaction between the variables.[10] However, even then, things are not always easy. Machine learning algorithms are focused on performance, not interpretability. While the field of interpretable machine learning is taking off, the challenges are still very much there. Perhaps the problems will be there forever, as, in many cases, the world is just too complicated to express in simple relationships. Machine learning's success might be attributed to its ability to extract complicated relationships from the data. Maybe we are just not smart enough to think beyond simple relationships consisting of only a few variables.

This has caused issues, even for big companies employing teams of experienced data scientists. Amazon, for example, had a team working on a tool that could automatically screen CVs for software development positions. It was leaked that the algorithm was

---

[10] I've also written about interpretable machine learning on the *Data Scientist,* for those of you who want to read more: thedatascientist.com/interpretable-machine-learning/

penalizing women, because the majority of the CVs it had been trained on were from male candidates.[11] This ended up being a PR nightmare for Amazon, who swiftly scrapped the project.

Finally, another issue with machine learning, which relates to the previous, is that machine learning does not understand causality. If an infant sees an object falling down, it can quickly extrapolate and understand there is a rule called "gravity." A machine learning algorithm would have to see many examples of this happening, and then it would associate a high probability (e.g., 0.99) to objects falling. This clearly doesn't make much sense. It's like Jian-Yang's app in the brilliantly acerbic sitcom *Silicon Valley*, which identifies food as either a "hotdog" or "not hotdog."

This is also exemplified by some recent advances in machine learning, such as ChatGPT and Large Language Models. These models can generate human-like responses to a wide range of queries and prompts, making them useful in a variety of applications, from customer service to creative writing. However, ChatGPT is unable to fully think in logical or causal terms. While the model can generate responses that appear to be logical or causal, it is not truly capable of understanding the underlying logic or causality of a given situation. This means that it may sometimes generate responses that are technically correct but don't actually make sense in the context of the conversation.

Judea Pearl, one of the prominent computer scientists in the history of AI, has heavily criticized machine learning's inability to understand causal relationships. For all the innovations that have gone into machine learning, it is still based on associative learning. The ladder of causation has three levels, and in order to build truly intelligent machines, Pearl argues, we need to move up from the first level (association) to the second and third levels, where causal reasoning actually resides. Without an understanding of causality, we don't have intelligent algorithms; we simply have "curve fitting machines." While some of the fits that these algorithms perform are impressive, achieving true artificial intelligence requires a paradigm shift in our way of thinking, one which places causality at the center.

---

[11] www.bbc.co.uk/news/technology-45809919

CHAPTER 9

Causality: Understanding the "Why"

"Shallow men believe in luck or in circumstance. Strong men believe in cause and effect."

—Ralph Waldo Emerson

Uncertainty appears in many forms. Most people associate the word "uncertainty" with the future, and rightly so. If we could predict what the stock market would look like tomorrow, we'd be overnight billionaires. If we can't predict that tomorrow it will rain, we might get soaked. However, a key part of uncertainty is understanding how events are *connected*. This question shows up again and again in science, as well as in everyday life.

In medicine, we need to understand which viruses might cause particular illnesses, and whether a specific substance is really a cure. That is, is a substance really the cause of a patient's recovery? In this context, looking for a cause means essentially identifying whether relationships in nature exist. The majority of scientists spend much of their time working on extracting this kind of relationship.

In history we are interested about the causes of significant events. What was the cause of World Wars I and II? What gave rise to certain ideologies? How did religions emerge and spread? This kind of causality relates mostly to understanding counterfactuals. When saying—for example, as we did earlier—that if Franz Ferdinand hadn't been murdered, then WWI wouldn't have happened, we attribute to the assassination the power of a cause. However, maybe WWI would have still occurred, but for other reasons. Investigating the causes of past, non-repeatable events is a very tricky and complex subject, which, naturally, leads to heated debates.

© Stylianos Kampakis 2023
S. Kampakis, *Predicting the Unknown*, https://doi.org/10.1007/978-1-4842-9505-2_9

In our social relationships (professional and personal), we are constantly trying to understand what caused particular "effects" or situations. Furthermore, we are actively trying to change our behavior, in order to reach desired outcomes. If we believe that dressing in a particular way will make us more likeable, then we might do that. If we believe that studying at a particular university will help us find a job, then we might apply to study there. If we believe that a particular action will make our boss give us a promotion, then we might go ahead and take this action. This kind of causal thinking relates closely to prediction. Similar to how machine learning algorithms work, we are extracting data from our world, and then we try to predict the future.

Understanding causality is not just a philosophical curiosity. Humanity has been very successful at surviving on this planet, partly because we are good at dealing with cause and effect. The problem of induction is, in many cases, essentially a problem of extracting causes and effects from nature. Also, understanding causality is not necessarily the same as building predictive models. There are fundamental differences between the two, which we'll consider now.

One of the things that every student of statistics learns is that correlation does not imply causation. Take a look at the scatterplot in Figure 9-1, which displays the relationship between two variables, x and y.

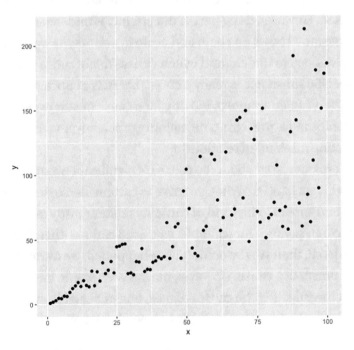

Figure 9-1. *Relationship between x and y*

Would you say there is a relationship between variables x and y? There seems to be. The relationship is not entirely clear, but as x grows, we can also see y grow. A measure of association between two variables is the famous correlation coefficient (first devised by Francis Galton). The correlation between those two variables, in this case, is around 0.8. The maximum score is 1, so this indicates a fairly strong relationship.

Now, let's take a look at the relationship depicted in the scatterplot in Figure 9-2. Would you say there is a strong relationship between the two variables? You bet there is. The correlation between x and y in this case is close to 1, indicating a nearly perfect linear relationship.

So, can we say that x is a cause of y in either of these two graphs? If yes, then can we say that it is more likely that x is a cause of y in the second graph? What do you think?

Well, what if I told you that in the first graph, x is the result of some national exam, like the SAT (in the United States) or the A levels (in the UK), and y is the average salary for the same person over the next 12 months; and in the second graph, x is the price of IBM's stock, from when it was released until 2000, and y is the percentage of people who drink Coca-Cola at least once per month in the state of Pennsylvania. Would you now say that one was the cause of the other?

While these examples might seem ridiculous, they prove the point: correlation does not imply causation.

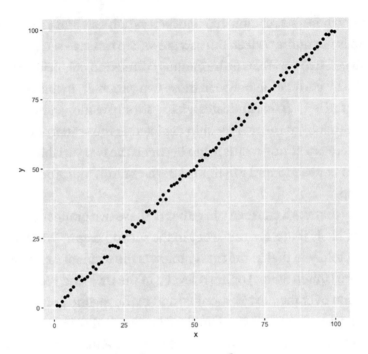

Figure 9-2. *Stronger relationship between x and y*

Probably no other source on the Internet has demonstrated this better than Tyler Vigen's website.[1] Tyler Vigen collects all kinds of amusing and irreverent correlations on his website. For example, in the graphs in Figure 9-3, you can see how the per capita cheese consumption correlates to the number of people who died by becoming entangled in their bedsheets. You can also see how the number of people who died by drowning in a pool correlates with the number of films that Nicolas Cage starred in.

---

[1] tylervigen.com

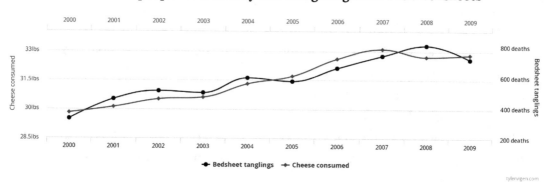

Figure 9-3. *Spurious correlations. Used under Creative Commons license*

These kinds of relationships are called "spurious correlations." The correlations between the two variables are indeed high, but there is no real relationship, causal or otherwise, between them.

So, what is the point of all this? We can conclude the following:

1) Correlation does not imply causation.

2) Tools used for association might not be good for understanding causation.

3) You can't always study causality within the context of a closed system. You might have to look outside of it.

The first point has, hopefully, become apparent by now. The second point is a very interesting and oft misunderstood one. Most of the data science models predict a variable *y* from one or more predictors *x*. However, the models could also be used to inverse the relationship. There is nothing to prevent us from predicting *x* from *y*. The model will work no matter what. The model itself can't help us understand whether *x* is causing *y*, or vice versa. This is true, from simple models, like linear regression, to complicated ones, like neural networks. If there is any causal relationship, it is assumed by the experimenter.

Which brings us to the third point. Causality within the context of most of our current models is something that has to be somehow assumed and controlled by the experimenter. If the models and our tools do not offer the capability to figure out the causal relationship, then it is down to the researcher to do that. The field of experimental design, revolutionized by Ronald Fisher, deals specifically with that.

This brings us to an interesting conclusion. While our current tools and methods are highly sophisticated, they still work on associations. Judea Pearl, who was awarded the Turing Award in 2011 for his contribution to computer science, has attacked current AI methods for their ignorance of causality. According to Pearl, identifying causal relationships is one of the hallmarks of intelligence. However, what our current methods are capable of is simply "curve fitting."

The key to unlocking intelligence in machines, Pearl believes, is related to them also understanding cause and effect. An interview in *Quanta* magazine elucidates how Judea Pearl sees the whole matter.[2]

> "Yet in your new book you describe yourself as an apostate in the AI community today. In what sense?"

> "In the sense that as soon as we developed tools that enabled machines to reason with uncertainty, I left the arena to pursue a more challenging task: reasoning with cause and effect. Many of my AI colleagues are still occupied with uncertainty. There are circles of research that continue to work on diagnosis without worrying about the causal aspects of the problem. All they want is to predict well and to diagnose well[...]"

> "I can give you an example. All the machine-learning work that we see today is conducted in diagnostic mode—say, labeling objects as "cat" or "tiger." They don't care about intervention; they just want to recognize an object and to predict how it's going to evolve in time."

> "I felt an apostate when I developed powerful tools for prediction and diagnosis knowing already that this is merely the tip of human intelligence. If we want machines to reason about interventions ("What if we ban cigarettes?") and introspection ("What if I had finished high school?"), we must invoke causal models. Associations are not enough—and this is a mathematical fact, not opinion."

---

[2] www.quantamagazine.org/to-build-truly-intelligent-machines-teach-them-cause-and-effect-20180515/

Causality is a demanding subject, and a proper treatment of it might be still eluding us. But how has science and philosophy treated this subject in the past?

How Do We Approach Causality?

"Because reasoning about causes and effects is a very difficult thing, and I believe the only judge of that can be God. We are already hard put to establish a relationship between such an obvious effect as a charred tree and the lightning bolt that set fire to it, so to trace sometimes endless chains of causes and effects seems to me as foolish as trying to build a tower that will touch the sky."

—Umberto Eco, *The Name of the Rose*

The first person to study causality was Aristotle. He had said that, "We think we have knowledge of a thing only when we have grasped its cause." According to Aristotle, there are four causes of things—the material cause, the formal cause, the efficient cause, and the final cause.

The *material cause* refers to the material out of which an object is composed. For example, the material cause of a bronze statue is the bronze metal. The matter is a cause for the statue, because it contains the potential of being turned into one.

The *formal cause* refers to the shape of an object. For example, the shape of a statue is its formal cause. So, a human body has a formal cause, in the sense that the different body parts are arranged in a particular way. A table is a table because it has four legs, otherwise it would fall down.

The *efficient cause* refers to the agent that has transformed the raw material into its form. The artisan who can take the bronze and turn it into a statue is the efficient cause of the statue.

The *final cause* plays the most important role in Aristotelian philosophy. The final cause is the teleological purpose of the object, from the Greek *telos*, meaning "end," "purpose," or "goal." In other words, the reason that it was made. A fork's reason of existence, for example, is to be used for eating.

Aristotle studied causes, but did not offer a framework for studying causality per se. This was actually the case for much of philosophy until modern methods of science appeared. This might be due to the fact that for most of history, the world was religious. Indeed, the teleological argument—advanced by, among others, William Paley—is the fifth of St Thomas Aquinas' "Five Proofs" for the existence of God. Hence, the cause of things was likely to have been attributed to supernatural and divine entities. The philosophy occasionalism exemplified this approach. According to occasionalism, there is no cause and effect. It is God who is causing all events. It's just that because events might take place one after another, we are under the illusion of causality.

David Hume also studied the topic of causality. According to Hume, there are two kinds of concepts: analytic and synthetic. Analytic concepts are ideas for which we can infer relationships simply by thinking about them. Mathematics is the prime example. Synthetic concepts are generated from our experience. Our mind is not capable of observing causality directly. It is a synthetic concept that we get from experience.

Instead of causal relationships, what we observe is simply one event following another. Hume set out three conditions that would determine whether an event A caused an effect B: A should always happen before B, and there should be proximity in terms of time and space between the two events. During Hume's time, the conversation around causality was directly related to the problem of induction and epistemological questions about the source of knowledge. Rene Descartes, among other continental rationalists, believed that we were born with innate ideas, whereas British empiricists—such as Hume, and John Locke—believed that we get our knowledge through our interaction with the world. The classic example of our mind being a *tabula rasa*—a blank slate—is attributed to Locke.

After the Renaissance, and as science progressed, divine explanations, as well as deep philosophical questions, gave rise to the laborious study of cause and effect in nature. Step-by-step, the sciences, especially the physical sciences, identified relationships between atoms and molecules, extrapolating laws and improving our understanding of the cosmos. Now we know, through the efforts of brain sciences, that the truth in the argument between empiricists and rationalists lies somewhere in the middle: We do have some innate knowledge, and we also learn from experience.

A good example of the pragmatic approach of the scientific method's search for causal patterns is "Koch's Postulates." Robert Hermann Koch (1843–1910) was a German physician who came up with four postulates that determine whether a species of bacteria can cause a disease:

- The bacteria must be present in every case of the disease.

- The bacteria must be isolated from the host with the disease and grown in pure culture.

- The specific disease must be reproduced when a pure culture of the bacteria is inoculated into a healthy susceptible host.

- The bacteria must be recoverable from the experimentally infected host.

The postulates were later found to be in need of adjustment, but they brought much clarity to the field of medicine.

Other important individuals in the study of causality are the statisticians Donald Rubin and Jerzy Neyman (whom you've met quite a few times in this book already). They came up with what is now called the "Neyman-Rubin Potential Outcomes Model." This model is based on counterfactual conditional statements. A counterfactual is a hypothesis about what *might* have happened. For example, let's say that you take two people who are exactly the same on all accounts. You give substance A to person A and leave person B untreated. If person A grows wings (or whatever other significant outcome you're after), then you assume that, since everything is the same between them, the wings must have grown as a result of the substance.

This a pretty common way to study cause and effect in a medical setting.[3] We split the samples into two groups: treatment and control. Both samples are equal in terms of variability. The treatment group is given a treatment, and the control group is given a placebo. Any effect within the treatment group is attributed to the treatment. Besides medicine, you can imagine that treatment here might also refer to: psychological therapy, educational courses, nutritional plans, and so on.

In spite of the progress of science in extracting all sorts of useful relationships from nature—which have translated to huge technological achievements—the discussion about studying causality in and of itself is fairly recent. While the goal of experimentation in science is to help us figure that out, in many cases the results can be confusing. Probably, there are no better examples to be found of this in studies that try to link medical outcomes with lifestyle factors.

For example, people who eat less meat might live longer. Hence, we might assume that red meat causes people to die earlier. However, some people might eat less meat because they are more educated about the current dietary trends. People who read more

---

[3] A full explanation of the model should also go into the terminology and the actual assumptions of the model, but these are beyond the scope of this book.

about the current dietary trends might be people with a higher educational and income level—arguably they can afford the time (and money) to search out plant-based organic produce. This means they might have a better quality of life and access to better doctors, which can explain the higher lifespan.

I am not saying that red meat is good or bad—in fact, the benefits and negatives of red meat are highly contentious—but there are plenty of similar studies that you can find in academic journals online. Simply replace "red meat" and "lifespan" with some other lifestyle habit (coffee, red wine, veganism) and a medical outcome, and you are guaranteed to find thousands of papers, sometimes conflicting each other.

The problem with identifying causal relationships is that things are rarely clear and easy. Quite often causes get mixed up, with potentially uncanny or comic effects. This aforementioned issue is called "confounding" and is a major consideration when studying causal relationships.

According to Judea Pearl, this is because the science of statistics has focused on the wrong thing for a long time,[4] possibly too long. However, it might also be because causality is a very difficult project to treat mathematically, and that many simpler methods have actually worked well for a long time.

Sometimes these methods can dramatically fail. One of the best examples of when this happens is the birthweight paradox, which is a classic confounding problem. The birthweight paradox is the observation (confirmed by multiple studies) that, for babies with low birthweight whose mothers smoke, the mortality rate is lower than for low birthweight babies whose mothers do *not* smoke. What does this mean? Traditionally, low weight at birth has been associated with higher mortality rates. Also, it has been confirmed that smoking is not good for the babies. However, if a baby has low birthweight *and* the mother smokes, instead of the baby having a higher mortality rate (as someone would expect), it is lower than if their mother didn't smoke.

This might sound like nonsense, but it is a confirmed phenomenon.[5] How is this explained? It is explained through confounding. A paper in the *International Journal of Epidemiology*[6] provides a possible explanation:

---

[4] Pearl, Judea (2018), *The Book of Why: The New Science of Cause and Effect.*

[5] Yerushalmy, J. (2014). The relationship of parents' cigarette smoking to outcome of pregnancy—implications as to the problem of inferring causation from observed associations. *International Journal of Epidemiology*, 43(5), 1355-1366.

[6] VanderWeele, T. J. (2014). "Commentary: Resolutions of the birthweight paradox: competing explanations and analytical insights." *International Journal of Epidemiology, 43*(5), 1368-1373.

"The intuition behind this explanation is that low birthweight might be due to a number of causes: one of these might be maternal smoking, another might be instances of malnutrition or a birth defect. If we consider the low birthweight infants whose mothers smoke, then it is likely that smoking is the cause of low birthweight. If we consider the low birthweight infants whose mothers do not smoke, then we know maternal smoking is ruled out as a cause for low birthweight, so that there must have been some other cause, possibly something such as malnutrition or a birth defect, the consequences of which for infant mortality are much worse."

So, if a mother is smoking, then this is likely the cause of the low birthweight. If a mother is not smoking, then the low birthweight is the result of some other, possibly more serious, condition. Hence, the lower mortality of babies with low weight in birth whose mothers smoke.

Another popular paradox is Simpson's paradox. A trend might appear when splitting the data in groups, but it disappears once the groups are combined. A famous example of this is the "UC Berkley Gender Bias" study. A paper in 1974[7] demonstrated that measuring bias might not be that easy. In the results of the paper, men were more likely to be admitted to Berkley than women, with 44 percent of the men applying ending up being admitted, and only 35 percent of the women being admitted.

However, when the applications were broken down per department, what was observed was that men were applying in less competitive departments, where admission rate is higher. Women applied in more competitive departments, with low admission rates overall. The actual data, broken down by department, displayed a small bias in *favor* of women.

Confounding and causality are tricky issues. In an ideal world, we would be able to study causality through counterfactual explanations, by splitting homogeneous samples in multiple groups, and providing a different treatment to each group. However, the real world is messy. Quite often we can't do this, which is why, in the study of social systems, debates are so common. A major issue in this is that our models cannot substitute for knowledge of the world. Quite often, we have to look outside the model and use our

---

[7] Bickel, P. J., Hammel, E. A., & O'Connell, J. W. (1975). "Sex bias in graduate admissions: Data from Berkeley." *Science*, 187(4175), 398-404.

common sense, in order to understand how variables are related. But as the birthweight paradox demonstrated, this is not always easy; in fact, multiple confounding models have been proposed to solve it.

However, this is what Judea Pearl set out to do. The latest attempt to deal with causality is Pearl's do-calculus. Pearl dedicated a large part of his career in causal reasoning, which also explains his passion and heated arguments with statisticians about the topic. According to his calculus,[8] there are three levels of studying causality (Pearl calls this the "ladder of causation"):

1) Association

2) Intervention

3) Counterfactuals

Association is related to seeing. The questions associated with seeing are "What is?" and "How would seeing X change my belief in Y?" For example, what does a symptom tell me about a disease, or a survey about an election result?

Intervention is associated with (you guessed it) intervening. You can ask "what if," or "what if I do X?". For example, will taking aspirin cure my headache?

The final level, counterfactuals, is associated with imagining and introspection. We ask the "why?," which is the most powerful question one can ask in philosophy and science. We can ponder and analyze whether it was X that caused Y, and what would have happened if we had acted differently. This captures many of the questions we set out in the beginning. If you observe that you took an aspirin and then you stopped having a headache, you can ask (in counterfactual reasoning) was it the aspirin that cured your headache? Would Kennedy have been alive 20 years later had Oswald not shot him? Would WWI have started if it were not for the assassination of Franz Ferdinand? And how would have my life been had I chosen to quit my job five years ago?

According to Judea Pearl, counterfactual reasoning goes back to the questions that David Hume and John Stuart Mill set out to answer. However, it is only now that we actually have the symbolic tools and methods to study these questions in a structured way. Counterfactual prediction is retrospective reasoning. By analyzing different versions of the past and how they compare to now, we can understand the "why."

---

[8] Pearl, J. (2018). "The Seven Tools of Causal Inference with Reflections on Machine Learning." Technical Report, Communications of Association for Computing Machinery.

Is Pearl's do-calculus the best way to treat causality? Time will tell, but what we can be certain of is that the study of causality will keep on playing a huge role in both science and our lives. In fact, reasoning about causal relationships is an inescapable part of life, closely related to the problems discussed in previous chapters, such as the problem of inductive reasoning. Our brains have, in fact, developed to an extent partly as a response to this problem.

Causality in Our Mind

"Look in my face; my name is Might-have-been; I am also called No-more, Too-late, Farewell."

—Dante Gabrielle Rossetti

While causality has been hotly debated by philosophers and mathematicians alike, nature has given us hints to how the problem might be solved. The term "theory of mind" refers to the cognitive ability of attributing causes and effects to someone else's state of mind. We all have mental states: beliefs, desires, motivations, and so on. The knowledge that we have those states and the knowledge that other individuals have those states and can act upon them defines the theory of mind.

While this might sound like a trivial observation for humans, it is far from it. It is one of the things that separates us from most animals (but not all) and it is a sign of intelligence. (In 2016, a study was published with potential proof that crows have a theory of mind. The crows seem to hide food as a response to them believing that other crows might try to steal it.[9] Similar observations have been made in primates.)

There are situations in which the theory of mind gets it completely wrong. In paranoid schizophrenia, for example, the person suffering from the condition might attribute completely irrational causes to the actions of other individuals. For instance, someone might assume that people are out to harm them, without any provocation or benefit, to themselves. Another case where theory of mind is disrupted is in people with autism; one of the defining characteristics of autism is that it can lead to difficulties in employing theory of mind and understanding or connecting to other people.

---

[9] Bugnyar, T., Reber, S. A., & Buckner, C. (2016). "Ravens attribute visual access to unseen competitors." *Nature Communications, 7,* 10506.

The theory of mind evolved as societies grew more complex, and understanding the motivations and beliefs of the people around us became of paramount importance to our survival.

The fact that evolution has gifted us with this ability clearly demonstrates the importance of social causal reasoning. However, the fact that we have multiple disciplines dealing with cognition and behavior, from psychology to neuroscience, demonstrates that our theory of mind might be incomplete. We still don't have a perfectly clear idea of what motivates individuals in every situation. Furthermore, we have created complex social systems that seem to have a soul of their own.

Many atrocities have also been committed based on false beliefs about humanity. Mao's cultural revolution, which led to the deaths of hundreds of thousands—possibly millions—of people, was based on wrong assumptions about human nature.[10] All political systems need to make assumptions about human nature, from the capitalist's driving belief that "people are basically selfish," to a belief commonly associated with the left—and first proposed by Jean-Jacques Rousseau—that humans are essentially benevolent and that society leads to corruption. The fact that we are still debating these topics demonstrates the complexity of figuring out what are the causes and effects in the world around us. Freud built his whole career on the idea of the subconscious, and he argued that we don't really know our *own* motivations in the first place.

Another way that causality enters into our mind is through counterfactual reasoning. Counterfactual reasoning can be split in two categories.[11] Upward and additive counterfactuals generate negative emotions, like regret. Their function is to prepare the individual to take a different course of action if a similar situation is encountered in the future. How many times have you said to yourself, "If only I had done X, then Y would have happened," or "If I had or hadn't done X, then maybe I would have avoided Y?" This is one kind of counterfactual thought. They're called upward counterfactuals, because they represent ways through which the actual event could have been better if, as the saying goes, life had been "on the up."

The other kind of counterfactual thoughts consists of downward and subtractive counterfactuals. These are the opposite of upward and additive counterfactuals. They represent counterfactual reasoning about events that could have gone worse. Downward

---

[10] Greene R. (2018), *The laws of human nature*

[11] Roese, N. J. (1997). "Counterfactual thinking." *Psychological Bulletin,*121, 133–148

counterfactuals generate positive feelings, such as "If I hadn't gone to college, then, I wouldn't have the nice job I have now," or, "If I hadn't met my partner, we wouldn't have had a family." These counterfactuals make us feel better about our experienced outcomes.

Not only have counterfactuals been found to have an overall positive effect on the individual, but they also play a huge role in us getting ready for the "real world." It's been argued that the extended period of development in humans, compared to other animals, is in part because of the need to develop improved causal cognition capabilities.[12] When children play, they explore the world, both physical and social. They construct *what-if* scenarios. During this exploration, they improve their cognitive skills, including, among them, counterfactual reasoning. It's been found that children even as young as 2.5 years old might possess the ability for counterfactual reasoning.[13]

While the importance of counterfactual reasoning might not be apparent at first, just think how often you come up with similar statements yourself. And just take the economic or political section of any newspaper; the conflicting opinions about the different actions that political leaders should have taken for the economy, or a series of other matters, is an example of counterfactual reasoning. It is so embedded in us that quite often we don't even realize we're using it.

Counterfactual reasoning can be seen as a tool against uncertainty, but also as a generator of uncertainty as well. When we think about what might have been, we're using a model of the world in order to train our mind to respond in a different or similar way in future situations. The knowledge you acquired after the fact is an additional data point that you can use to extract information about how the world works going forward.

However, counterfactual reasoning itself assumes that we have a correct model of the world. The reality is that we will never have a 100 percent accurate model of the world in our mind. The counterfactual proposition, "If I had an extra tire in the trunk, then the flat tire wouldn't have delayed me as much," is a good example of a sound piece of advice you can give to yourself—it's retroactive reasoning. However, the counterfactual,

---

[12] Buchsbaum, D., Bridgers, S., Skolnick Weisberg, D., & Gopnik, A. (2012). "The power of possibility: Causal learning, counterfactual reasoning, and pretend play." *Philosophical Transactions of the Royal Society B: Biological Sciences,* 367(1599), 2202-2212.

[13] Harris P. L.& Kavanaugh R. D. (1993), "Young children's understanding of pretense." *Monogr. Soc. Res. Child Dev.* 58, 1–107.

"I shouldn't have bet on Manchester United to have won the Premier League last season," after the league has finished might not be very useful, unless this proposition can be accompanied by other criticisms and reasoning about the correct way to bet on sporting outcomes.

This example provides the close link between many of the things we've seen so far. Counterfactual reasoning that goes beyond trivial common sense (e.g., "I shouldn't have drunk ten pints of beer the night before an important meeting.") or the physical world ("If I would have been servicing my car more often than once every five years, it wouldn't have broken down.") is fundamentally complex. It requires a model of the world, but also a prediction about something in the past. Its usefulness lies in assuming that, in the future, we might encounter a situation like the one we've encountered before. It is a true hallmark of intelligence, and one that we might have to encode into machines if we ever want to create *real* intelligence.

However, we wouldn't need to worry so much about what the past *would* have been like if we could just predict the future. Maybe we can find a way to actually extract all the relevant information we need from the past and use it in order to predict the future. A retail store is not so much interested in how it could have beaten its competitors two years ago if the manager had taken different actions. It is more interested in predicting new trends and demand for products. We're not interested in how our picnic would have gone if we had checked the weather forecast and we had taken a couple of umbrellas with us. Rather, we're interested to know what the weather will be tomorrow.

And, finally, many people have said that if they had bought Google or Apple stocks decades ago, they would be rich by now. However, in one or two decades, some other people will be complaining over their poor ability to predict that some other companies, which were maybe incorporated last month, will be worth billions by then. Hindsight is powerful and very seductive—but even more powerful is the ability to predict the future.

CHAPTER 10

Forecasting and Predicting the Future: The Fox and the Trump

"Those who have knowledge, don't predict. Those who predict, don't have knowledge."

—Lao Tzu

"I always avoid prophesying beforehand because it is much better to prophesy after the event has already taken place."

—Winston Churchill

Forecasting has been described as a "dark art." There is something inherently beguiling about it that latches on to popular imagination; from all the applications of data science, forecasting is among the ones that fascinate the public the most. Predicting the future is something that has always excited humans, and for obvious reasons. We could argue that the main goals of science are explaining causes and building models that help us predict and control the world around us. For the longest part of human history, predicting what will happen in the future was associated with witchcraft, deities, and other supernatural entities—not any more.

Forecasting in data science has become more or less synonymous with time series prediction. A "time series" is simply a set of values in time. Examples of time series are stock market prices, prices and demands of goods and services, and temperatures.

Forecasting is based on a simple idea: The future is going to look like the past, hence, we can use information from the past to predict the future.

© Stylianos Kampakis 2023
S. Kampakis, *Predicting the Unknown*, https://doi.org/10.1007/978-1-4842-9505-2_10

There are many types of forecasting models, but for the purposes of this book, a high-level breakdown creates two main categories:

1) Endogenous models

2) Models that take into account exogenous variables

These definitions might not be encountered in the academic literature as they are presented here, but they will help you understand the different kinds of information used in models, and the different kinds of assumptions made.

The first category of forecasting techniques (endogenous models) describes what most people associate with traditional forecasting; it's based on two main assumptions:

1) The future is going to look like the past.

2) The information provided in the time series is sufficient to predict the future.

If you think about it, these assumptions are both reasonable and wrong, at the same time. The future is definitely going to be influenced by the past. If I asked you what the temperature is going to be in the next hour, you would give me a prediction that wouldn't be too far off from the temperature right now. However, how far do we have to look into the past, and how far can we predict into the future? Does the temperature from last week influence the temperature in the next hour? And can we predict the temperature tomorrow? What about next week? Or next month?

It is clear that for some time series, the aforementioned approach might make more sense than for others. Let's say that you want to predict the demand for a particular product—winter gloves. Common sense dictates that the demand for this product will be higher during the winter and autumn months. Given that the market doesn't change much (no new competitors show up with cheaper products, for example), you might expect demand to look very similar each year. Likewise, we might observe trends over many years. A famous time series example is one of air passengers after World War II, when a booming economy and industry led to a constantly rising number of passengers every year.

The plot in Figure 10-1 is produced by R, using the AirPassengers dataset, which was originally created by Box and Jenkins and describes the monthly totals of international airline passengers from 1949 to 1960. You can see that there are some stable trends, like seasonality, and a growth pattern over the years. This is a type of time series with many regularities that we can successfully model using the endogenous approach.

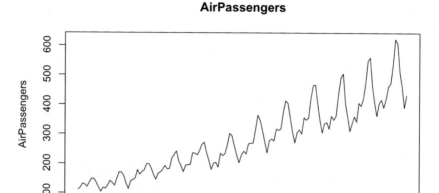

Figure 10-1. *Classic example of a time series*

But let's say that something changes regarding the winter glove example. A new competitor makes a cheaper alternative. There is a new fashion trend. Global warming means that winters are warmer, so people are buying fewer gloves. In that case, the model will stop working. Some of those changes, for example changes in temperature, might happen gradually over time. So the model can detect them. But in many cases, they can happen abruptly and catch us off guard.

Another good example comes from financial time series and retail. Let's say that you have a model to forecast demand for a product or for stock market prices. Then a crisis happens (like the one that took place in 2008), which causes a global recession, with the government intervening with new regulations in order to rebalance the economy. A similar situation took place during the COVID-19 pandemic. This is another event that might have been difficult to foresee, and completely changes the underlying system we're modeling. We can hardly expect the time series we're studying to contain enough information about what kind of regulations a government will impose!

In Figure 10-2, you can clearly see recessions in the shaded areas. The recession after 2008 was unprecedented in recent history, both in terms of its impact and its duration.

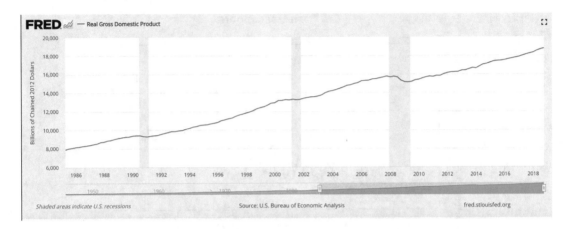

Figure 10-2. *GDP of the USA from 1986 until 2019. Source: U.S. Bureau of Economic Analysis, Real Gross Domestic Product [GDPC1], retrieved from FRED, Federal Reserve Bank of St. Louis;* `fred.stlouisfed.org/series/GDPC1`

This kind of sudden change in a system might render the model completely useless. This is called *concept drift*. It is difficult to detect until well after the event has taken place and has been the bane of many forecasters.

Another criticism of endogenous models comes from the fact that, often, there is information that can be helpful for forecasting a time series—perhaps some other time series—that isn't taken into account. We would expect, for example, that the GDP and the inflation rate might affect demand. Both of these time series might contain information that is not directly contained in the time series we are predicting. Or, let's say that you are trying to forecast the yield of crops. Climate and soil information would be immensely useful.

For that purpose, we developed models that can include information from exogenous factors as well. These models were developed much later than endogenous models. One of the main drivers of these models was the advent of the computer, which allows forecasters to calculate the results of complicated models in seconds.

There are many types of categorizations of forecasting models with exogenous factors. Again, for the purposes of this book, I believe there are two main categories: adaptations of endogenous models and general-purpose models.

By adaptations of original models, I am referring to extensions of endogenous forecasting models to also include exogenous variables. One such model is ARIMAX. The original endogenous model is called ARIMA (Autoregressive Integrated Moving Average). ARIMA is a pure endogenous model, using past values of the time series in order to predict the future. ARIMAX can take into account other variables, not related to the time series. Again, coming back to the previous examples, we might want to predict demand for a product based on past demand (the endogenous component), plus the GDP and the inflation rate of a country.

Regarding the general-purpose models, these are models that can be used for other purposes,[1] but they can also be applied to time series. That's the case with many machine learning models. The most classic examples of models applied to time series forecasting are neural networks and Gaussian processes. There are also more advanced approaches that combine elements of machine learning and more traditional techniques, like Bayesian structural time series models, which are favored by Google.[2]

General-purpose models have gained more ground in recent decades, due to their flexibility and ability to incorporate multiple factors into the predictions. However, this doesn't mean that they are always better. Don't forget the bias-variance trade-off. Often, simple models can be more powerful than advanced ones. Soon you'll see that deciding the "best" method is a major debate in the forecasting community. But first...

A Brief History of Forecasting

"I have seen the future and it is very much like the present, only longer."

—Kahlil Gibran, *The Prophet*, 1923

The history of modern forecasting has to start with Udny Yule (1871–1951) and Eugen Slutsky. Udny Yule was a British statistician, with multiple contributions to the science of statistics. He wrote *An Introduction to the Theory of Statistics*, one of the most popular textbooks of the time, in 1911. He was the first person to discover the Yule Paradox (now called Simpson's paradox), a phenomenon where a trend that exists in groups disappears when the groups are combined.[3] He also made contributions to genetics, epidemiology, and time series.

Eugen Slutsky (1880–1948) was a Soviet mathematician who contributed to economics and probability theory. Slutsky's greatest contribution to economics and statistics is now known as the Slutsky-Yule theorem.

---

[1] For example, neural networks can be used for both regression and classification.
[2] ai.google/research/pubs/pub41854
[3] www.britannica.com/topic/Simpsons-paradox

The two men didn't know each other, but they reached the same conclusion at the same time: That the moving summation or average of a random series may generate oscillations when no such movements exist in the original data. In other words, you can get a regular periodic pattern from completely random data. Both Slutsky and Yule published their papers in 1927.

Slutsky used an unconventional method for that time, which was a forerunner of modern computer simulations. He and his colleagues drew random numbers from a lottery, and then added up the results of the last ten numbers. He realized that this created patterns similar to the ones observed in British business cycles. Similar patterns are also observed in nature, in phenomena like waves. The discovery that completely random numbers can cause regular patterns was ground-breaking at the time. Translated to economic theory, this meant that the economy could experience shocks without any exogenous factors influencing it.

This was an important discovery for economics, but also for time series analysis, as this was effectively the first autoregressive process that was used. An autoregressive process is simply a model that predicts the future based on a weighted average of the previous values. The past contains the information we need to predict the future.

Things moved on after that. Charles Holt (1921–2010) came up in 1957 with the first modern forecasting method, called "exponential smoothing." Whereas similar principles had been applied by Siméon Denis Poisson in the 17th century and adopted by the signal processing community, this was the first time that this kind of principle had been applied in time series.

Holt's method is based on a simple idea. The next value can be predicted by the previous values, with the weights of those values degrading exponentially over time. This means that a prediction considers all of the history, but for older datapoints, the weight is practically zero. Figure 10-3 illustrates this more clearly. Let's say that you want to predict what the demand of a product will look like. You're using daily data. The weights for an exponential smoothing model would look like the ones in Figure 10-3. The previous day has the largest weight of around 0.35. Days more than a week into the past get practically 0 weight.

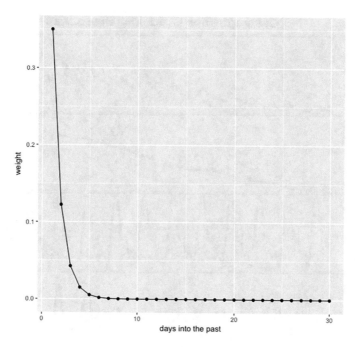

Figure 10-3. *Example of weights of an exponential smoothing model*

This is a method that works well in practice, but obviously it has its limitations. It does not take into account, for example, seasonal patterns. That's why more extensions to this method came later, such as Holt-Winters smoothing and Brown's double exponential smoothing, among others.

A major contribution to the field of forecasting came from George Box (1919–2013) and Gwilym Jenkins (1932–1982). They discovered the ARIMA method, mentioned earlier. Their model is another application of the idea of using past values in order to predict future ones. However, it uses a different method to arrive at the weights. For a long time, ARIMA was considered state of the art in time series forecasting.

Another technique that came in around the same period was time series decomposition. Time series decomposition breaks down a time series into different elements. An example of time series decomposition is shown in Figure 10-4. You can see that the AirPassengers data, which you saw earlier, has a seasonal component, a trend, and random noise, which in this case seems to follow some kind of periodic pattern as well. This technique is used to better understand how a time series behaves.

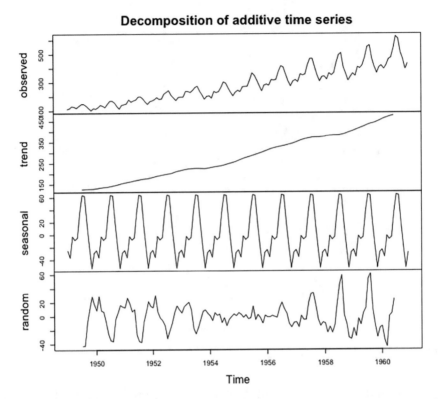

Figure 10-4. *Example of time series decomposition for the AirPassengers data. Adding all three components together (random, seasonal and trend) will give us the original time series*

Modern developments in forecasting have been inspired by machine learning and the availability of multiple different data sources and cheap computational power. Facebook, for example, released a tool called Prophet[4] which uses multiple different techniques simultaneously to automatically come up with a good forecasting model.

Also, another popular approach is to use general-purpose algorithms, such as deep neural networks, and feed them with multiple different sources of information, from time series to posts on social media.

Given the usefulness of forecasting, we are certain to start seeing more techniques coming out in the near future.

---

[4] facebook.github.io/prophet/

Forecasting in practice: Newton and the madness of men, Trump, Brexit, and losing money through mathematical modelling

"An unsophisticated forecaster uses statistics as a drunken man uses lamp-posts—for support rather than for illumination."

—Andrew Lang

Forecasting is a very useful application of data science, that much is clear. The previous section hopefully helped you understand how the problem is approached from a mathematical perspective. But how well does forecasting fare in practice? There are many successful applications of it across myriad industries.

But let's take a look at what happens when forecasting goes wrong.

As you can see, many of the forecasting techniques are just different versions of the same idea that the past can be used to predict the future. We mentioned earlier why this doesn't always hold true.

Our topic here is, of course, uncertainty—so how does a forecasting model deal it?

Many statistical methods of forecasting can explicitly model uncertainty, through the use of confidence intervals. That's the case for the ARIMA model, for example, which we mentioned earlier.

In Figure 10-5, you can see an ARIMA model that was fitted on the AirPassengers data, which you saw in Figure 10-1. The original AirPassengers data goes up until 1960. The predictions go up until 1975.

The shaded areas around the predictions are the confidence intervals at the 80 percent and 95 percent levels. A 95 percent confidence interval can be interpreted as the value of the forecast lying anywhere within this range, 95 percent of the time.

There is a clear pattern that emerges in this image. As years go by, the confidence intervals get wider and wider. After a point, the confidence interval becomes too large. At that point, the prediction stops being useful. If, for example, we used this model to predict the demands on airlines and also forecast revenue, the end numbers will look very different if we have 500 million passengers or 1 billion.

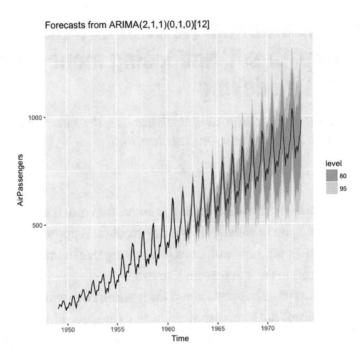

Figure 10-5. *Predictions along with confidence intervals for an ARIMA model used to predict future numbers of air passengers*

It's clear that even when the trends are relatively stable, the uncertainty grows over time, until it reaches a point where the model stops being applicable. But that's the optimistic scenario. In real life, we deal with more complicated systems, with many interconnected components. This can cause issues, usually in unexpected and negative ways. The field of finance exemplifies this better than any other.

A famous example of a very smart person losing a lot of money in the stock market is Sir Isaac Newton's story of losing a fortune in the South Sea Bubble. The South Sea Bubble was one of the first and biggest financial crashes of all time. In 1720, the House of Lords passed the South Sea bill. This gave to the South Sea company a monopoly in trade with South America. In exchange, the company lent the government £7million to finance the war against France—today that figure equates to a sum just shy of £100million, so you can see the appeal. Shares rose rapidly in value, with speculators and the public investing all their money in the hopes of excessive returns. The share's price went from £128 to over £1000.[5] Obviously, as it happens with these stories, the stock crashed, destroying fortunes.

---

[5] www.britannica.com/event/South-Sea-Bubble

Newton invested in the South Sea company and pocketed a profit of £7000. He decided to reinvest, taking a loss of £20,000 by the end of the bubble. This is the equivalent of around £3 million in today's money.[6]

Newton famously proclaimed that he "could calculate the motions of the heavenly bodies, but not the madness of the people."

While Newton didn't use any mathematical forecasting techniques, this story exemplifies the difficulty of making accurate predictions about the future when the system in question is exposed to multiple interacting factors. One doesn't have to look far to find similar stories in the recent past (the dot-com bubble and the bitcoin bubble being just two), but we also don't have to look far to find similar stories that *did* use sophisticated techniques.

One of the challenges with forecasting is that the accuracy of predictions in the future do not only depend on our model, but also on the stability of the underlying system. There couldn't be a better example of how this can go wrong than the case of Black, Scholes, and Merton. These three men are behind the famous Black-Scholes model, a formula for option pricing. It was introduced in 1973 and is perhaps the most famous model in finance. Scholes and Merton received a Nobel prize in economics in 1997.[7]

Scholes and Merton were seated on the board of the Long-Term Capital Management L. M. firm. The fund was initially successful, with returns between 20 percent and 40 percent, until it crashed in 1998, losing more than $4.6 billion in four months and requiring a bailout from the Federal Reserve. As a result, the theories of Merton and Scholes were discredited and attacked. Merrill Lynch reported that mathematical models "may provide a greater sense of security than warranted; therefore, reliance on these models should be limited."[8]

So, why did this happened? The models employed by Long-Term Capital Management L.M. were based on the efficient market hypothesis. This assumes that the investors are rational, and the market includes all relevant information in its pricing

---

[6] Benjamin Graham (2003), *The Intelligent Investor*

[7] Black had already passed away. Nobel prizes cannot be given posthumously, but the Nobel committee recognized his role.

[8] Roger Lowenstein (2000), *When Genius Failed: The Rise and Fall of Long-Term Capital Management*

decisions. The models didn't take into account extreme events that can potentially disrupt the market. We could get into a long discussion about this, and it is still a matter of dispute, but in short, the assumptions of their model did not hold true in reality.

This story exemplifies how extreme faith in models can have catastrophic consequences, and it was a forerunner of the events that unfolded just a decade later during the financial crash of 2008. Paul Wilmott and David Orrell discuss this in their book, *The Money Formula*.

One of the most popular tools in finance is the Modern Portfolio Theory. Pioneered by Harry Markowitz in the 1950s, this theory describes how a portfolio manager can allocate assets in a way that balances risk with returns. The problem with Modern Portfolio Theory is the same as the Black-Scholes formula. The models are correct as long as the underlying assumptions are correct. Plus, some of the mathematical manipulations of the Modern Portfolio Theory can drastically increase the uncertainty of the model. Wilmott and Orrell explain:

> "When we estimate returns, we are making a prediction about the future; but when we estimate risk, we are predicting the uncertainty in our forecast—a prediction about our prediction— which is even more difficult."

In finance, much like in other fields, scientists have tried to come up with all sorts of simplifying theories to explain uncertainty away using mathematical tools. However, the dynamic nature of the finance system makes these models susceptible to catastrophic failure.

Given that it is difficult to abstract the complexity of some systems into a formula, is there some easier way to make predictions? What if there was a structured way to ask people, since most of this knowledge that can't be easily encoded lies in their heads?

This is the principle behind the "wisdom of the crowds" method and prediction markets. According to this approach, a solid way to make forecasts about the future is by asking a large number of people and then averaging their predictions.

A more structured way to do this is through a prediction market. This approach came back into fashion again through the rise of blockchain. Examples of blockchain-based prediction markets include Augur[9] and Stox.[10] The main concept in a prediction market is that you ask a prediction about something, and then people give their own guesses. Successful predictions are rewarded.

---

[9] www.augur.net/

[10] www.stox.com/

Prediction markets can be seen as a generalized version of traditional betting. Whereas in betting, the available markets are determined by the house, prediction markets are more democratic and more flexible. For example, you can ask for a prediction of a real number (e.g., the total rainfall over a geographical region, or the price of a stock) around any arbitrary matter.

Obviously, another source of information that forecasters have used (and are still using) is information from betting sites, since the principle is the same. When enough experts get together to decide about something, and when there is money at stake (to ensure that the participants will put enough thought into the problem), then we can expect to get good forecasts.

Has this worked? Well, yes and no. Odds do provide a useful source of information. However, the biggest forecasting disasters in the 21st century prove just how catastrophically wrong predictions can get: The misprediction of the American 2016 election and the Brexit vote. In both cases, the majority of forecasters completely missed both outcomes, instead believing with high confidence that Hilary Clinton would win the election and that the United Kingdom would choose to stay in the European Union.[11] So, what exactly were the odds predicting? Until the very last minute, the odds were predicting that Hilary would win.[12] The polls didn't do much better. [13]

So, what went wrong there? Well, quite a few things. First of all, while betting odds can be a useful source of information, they are—of course—not 100 percent right. Secondly, polls in both cases suffered from serious biases. Many of the people who voted for Trump or Brexit weren't participating in polls, therefore giving a completely skewed picture as to the true number of supporters.

Even Nate Silver, who had risen to stardom when in 2008 he correctly predicted the winner of all 50 states, got the prediction wholly wrong. His site `FiveThirtyEight` gave Trump only a 28.6 percent chance of winning. After that, Silver came up with a series of posts explaining what went wrong and what can be fixed, but there were two fundamental problems.

The first one, which I just touched on, was the fact that the polls were biased. This was a kind of epistemic uncertainty. However, it is difficult to know that this kind of bias exists until it is too late. Epistemic uncertainty is reduced when we get new methods or new tools to collect more data, more accurately. Indeed, it was social media that

---

[11] For more details about this, go to thedatascientist.com/election-forecasting/.
[12] electionbettingodds.com/
[13] ig.ft.com/sites/brexit-polling/

contained information that was not accessible by polls. Trump had lots of engagement on social media, and many of his supporters participated on social media, but never participated on the polls. For example, in some trend analysis, it's clear that Trump had more positive sentiment throughout the campaign.[14]

The second issue around predictions for events like elections is that those events are non-repeatable. An election takes place only once and that's it—with the exception of referenda, the Irish vote on the Lisbon Treaty being a recent example (EU officials forced a second vote because Ireland got it "wrong"). The only realistic interpretation of probability is the Bayesian interpretation. The Bayesian interpretation describes probability as a belief. So, saying that Hillary has 60 percent probability of winning doesn't mean that Hilary will win. It means that it is more *likely* that she will win. But it is still possible that Trump might win. Probability in this case gives us a confidence score, but we will never be completely certain about the outcome.

This is one of the fundamental issues with the Bayesian approach. Bayes provides an interesting interpretation of probability, and it also is very useful when we have prior knowledge about something. But at the end of the day, when we are building models to make predictions, we are basing everything on past data. We hope that at least some of the features of the current system we're exploring will be similar to those in the past.

For example, we might assume that polls provide an accurate representation of the opinions of the voters. We might include the national GDP into our model, assuming that there is a relationship between GDP and some political party. We might use information about trends, or seasonal phenomena. However, all these imply a notion of repeatability, in one way or another. So, elections on the one hand are unique phenomena (the exact same election will never happen twice), but on the other hand they are repeatable (American elections happen every four years, and in other countries they take place at regular intervals). The first point provides justification for the Bayesian use of probability, whereas the second one for a frequentist interpretation.

Frequentism works well in cases where the actual experiments we're running are 100 percent repeatable—for example, when studying physical and biological phenomena. However, we have seen a surge in the popularity of Bayesianism in the recent years, which is often promoted as the cure-all.

---

[14] techcrunch.com/2016/11/10/social-media-did-a-better-job-at-predicting-trumps-win-than-the-polls

Nate Silver wrote a whole book advocating Bayes.[15] However, I will be more cautious about taking sides in this old academic debate. Maybe that's partly because I am more applications-oriented, and maybe partly because no one has yet come up with a good theory to merge the two approaches. I am not sure if this is possible, and I am not sure what is the best way to approach this problem, but if we were able to do this, maybe we would be able to come up with more powerful models.

So, the next question is, which forecasting model is the best?

There has always been a debate in data science as to which algorithm is the best for a particular task. You learned earlier in the book about the No Free Lunch theorem and what this means. If we were to compare all algorithms across all possible prediction tasks in the history of tasks, then they would all be tied. However, practically speaking, our world is not completely random. We expect certain regularities in nature, and these regularities mean that some models might perform better than others on average.

The best forecasting method is obviously a hotly contested topic. There are two reasons for this. The first one is that, as it must have become apparent in this section, forecasting holds immense value for many problems in myriad industries all around the world. The stakes are extremely high. Secondly, forecasting in practice is riddled with issues. As some of the stories demonstrated earlier, quite often we don't know that we made a mistake in forecasting until it's too late.

The first large-scale comparison of forecasting methods was held in 1974[16] by Paul Newbold and Clive Granger.[17] This was not an open competition, but rather a comparison of different methods on 106 time series. One of the key results was that combining methods could result in better performance over a single method.[18]

---

[15] en.wikipedia.org/wiki/The_Signal_and_the_Noise

[16] robjhyndman.com/hyndsight/forecasting-competitions/

[17] www.jstor.org/stable/2344546

[18] Combining models is common practice today in many machine learning competitions, and there is sound theoretical justification as to why this is a good practice. It is also one of the reasons that machine learning pipelines can be complicated and computationally expensive, which makes many people (even in the world of tech) treat machine learning with a combination of awe and curiosity. Also, it is one of the reasons that data scientists command high salaries. However, in practice, machine learning often resembles more of a craft, rather than a science, as it quite often involves trying different things until you get something that works.

Spyros Makridakis and Michèle Hibon performed a similar experiment in 1979. They compared different methods on 111 time series and published their results.[19] This caused quite a stir in the academic community, with many academics disagreeing with the results. One of the main disagreements was whether more complicated methods would fare better than simple ones. Another point of disagreement was if a combination of methods could perform better than a single model. One of the widespread beliefs, at that time, was that there is some kind of true model waiting to be discovered for each time series. Combining models looked a bit like alchemy. Because of that, Makridakis and Hibon followed up with the first open forecasting competition.

The first M-competition (M for Makridakis) was held in 1982 and involved 1001 time series. The competition confirmed their results from the 1979 test. Put concisely, the core conclusion was that more complex methods do not necessarily perform better than simpler ones, and that combining models seems to be an effective way of improving performance.

The M-competition was followed by three more competitions (in 1993, 2000, and 2018). The last competition (M-4 competition) used a total of 100,000 time series. All competitions validated the learnings from the first competition. Simpler models seem to work best. There were not many submissions that used machine learning models, and in separate competitions, neural networks (which is one of the most popular models for forecasting) did not fare well.

So, what does this mean to everyone who is not a data scientist? Forecasting is difficult, and more complicated methods seem to confuse the noise with the signal. The history of the M-competitions and the debate around them demonstrates very clearly some of the challenges that forecasters face, and some of the fallacies they have committed over the years.

One of the biggest critics of the results published by Makridakis and Hibon in 1979 was Professor Maurice Priestley (1933–2013) from the University of Manchester. Priestley commented on the results, saying that[20]:

---

[19] www.jstor.org/stable/2345077

[20] Spyros Makridakis, Michele Hibon, and Claus Moser. *Journal of the Royal Statistical Society. Series A (General)*, Vol. 142, No. 2 (1979), pp. 97-145.

"The performance of any particular technique when applied to a particular series depends essentially on (a) the model which the series obeys; (b) our ability to identify and fit this model correctly and (c) the criterion chosen to measure the forecasting accuracy."

This essentially means that there is one true model waiting to be discovered. Think about the implications of this philosophy for a second. What Priestley was describing was a popular view among many of his peers at that time. That a time series, irrespective of its origin, can be predicted to a high level of accuracy, and the only thing separating us from that amazing model is discovering the true process that generates it. So, it doesn't matter whether this time series is the GDP of an economy, or the demand for a product, or the total number of traffic accidents—there is some kind of formula that nature obeys and we can approximate.

This belief was founded on two wrong assumptions. First, there is no guarantee that there exists a perfect model for a time series. Why not? Because we haven't found it yet. Secondly, there is no guarantee that this model is something that looks "nice" and neat in the traditional mathematical sense.

It's an assumption that has showed up many times before in science. There is some omnipotent solution to our problem, and this solution follows some form understandable by humans—in other words, it is effable and *we will find it*. A combination of models might not seem as beautiful, from a mathematical perspective, as a single model, but then again, why should it be? Either due to human hubris, or due to an unshakeable faith in our powers of reasoning, scientists often assume this kind of magic formula must exist. But, as you will see, this is far from necessarily the case.

One of the successes of the approach championed by machine learning has been the fact that experiment and measurement have replaced, to a large extent, theoretical considerations. Fast experimentation has led to the creation of new powerful methods and techniques for prediction. This has the downside that the models can at times look opaque, plus machine learning looks a lot like a craft, rather than carefully constructed science. But the approach of "try things first, come up with theory later" has worked well for the big data era.

Interestingly enough, the approach of combining models has even become the standard in weather forecasting. While we have quite accurate physical models of the weather, combining many of them gives a more accurate prediction than each individual model.

You've probably heard of the "butterfly effect." Edward Lorenz (1917–2008), the father of chaos theory, asked in a conference "Can a butterfly flapping its wings in Brazil set off a tornado in Texas?". While a butterfly probably lacks the power to do this, this example demonstrated how small changes in a complex system can have disproportionate effects. Lorenz was a meteorologist when he realized that small changes in the initial conditions of a simulation resulted in radically different results. Just as with global warming—we might welcome our Spring months being a couple of degrees warmer, but that slight increase melts ice caps. We can't know every detail about every phenomenon that can affect the weather. Hence, aleatoric uncertainty imposes a limit on what we can know.[21] But by running multiple models and using different initial conditions, we can get a better idea of the uncertainty of our prediction and of the range of possible results.

The results of the M-competitions seem to corroborate this story. Many interesting systems (including the real economy) are so complicated that we have no choice other than to accept the fact that we are facing aleatoric uncertainty. Only in a science fiction scenario can we imagine that we have the capability to collect such detailed data that the only thing separating us from the perfect prediction is just epistemic uncertainty. Experience has shown that reality imposes limits on how much we can fight off uncertainty—at least in practice.

How to Do Forecasting: A Story of Foxes and Hedgehogs

"The fox knows many things, but the hedgehog knows one big thing."

—Isaiah Berlin, *The Hedgehog & The Fox: An Essay on Tolstoy's View of History*

Forecasting is difficult—that much we do know. Interestingly, this is something that has still not become apparent to many professionals. A study by Prakash Loungani at the International Monetary Fund, based on a sample of companies from 63 countries, revealed that private sector forecasters were able to predict only 2 of the 60 recessions

---

[21] In theory, someone could argue that this kind of uncertainty is epistemic and we will be able to remove it with new measurement instruments. In practice, there is always going to be even a small amount of noise that amounts to aleatoric uncertainty.

that occurred.[22] Professor Philip Tetlock initiated a famous study in 1984, called the Expert Political Judgment study,[23] in which he asked 284 experts (academics, pundits, economists, advisers, etc.) to make thousands of predictions about topics like the economy, elections, important global events, and so on. By 2004, when the study finished, he had collected 28,000 predictions. The results demonstrated that the "average expert" had no more foresight than a chimp throwing darts at a chart, or a person making random guesses.

In a 2014 video,[24] Tetlock summarized his findings.

> "There was systematic overconfidence. Moreover, political analysts were disinclined to change their minds when they get it wrong. When they made strong predictions that something was going to happen and it didn't, they were inclined to argue something along the lines of, 'Well, I predicted that the Soviet Union would continue and it would have if the coup plotters against Gorbachev had been more organized,' or 'I predicted that Canada would disintegrate or Nigeria would disintegrate, but it's just a matter of time before it disappears,' or 'I predicted that the Dow would be down 36,000 by the year 2000 and it's going to get there eventually, but it will just take a bit longer.'
>
> So, we found three basic things: many pundits were hard-pressed to do better than chance, were overconfident, and were reluctant to change their minds in response to new evidence. That combination doesn't exactly make for a flattering portrait of the punditocracy."

It looks like many experts really *are* using statistics in the same way a drunk person uses a lamppost.

However, not all is lost. While it's easy to be critical of forecasters, what is more difficult is to provide constructive criticism and understand what can be improved. The world is uncertain, but not random. If the history of the study of uncertainty has proven anything, it is that when you can't predict something directly, try to take a different route.

---

[22] Loungani, P. (2001). How accurate are private sector forecasts? Cross-country evidence from consensus forecasts of output growth. *International Journal of Forecasting, 17*(3), 419-432.

[23] P. E. Tetlock (2006), Expert Political Judgment: How Good Is It? How Can We Know?

[24] www.edge.org/conversation/how-to-win-at-forecasting

We can't predict the result of the next roll of a die, but we can predict the average result of a large number of rolls. We can't predict the height of the next person we encounter, but we can get a 99 percent confidence interval of where the height lies.

So, what can we do? First of all, we need to separate qualitative from quantitative forecasting. Philip Tetlock has dealt with qualitative forecasting. That is, forecasts about events that are not based on some mathematical model. The forecasters might be taking into account data, but their judgments are not based on a mathematical model. Qualitative forecasting is better suited for questions that are not easy to put into a mathematical model. For example, predicting whether a war will break out in the next ten years requires a good understanding of international relations. Encoding this in a mathematical model is possible, but it's an overtly complex task. Hence, expert human judgement is a more practical approach in those cases.

Quantitative forecasting, on the other hand, is based on mathematical models. Quantitative forecasting can deal with levels of granularity or prediction that are not easily addressable by human judgment. For example, making an accurate prediction about the weather, down to a decimal point, while using tens of variables, is something better suited to a mathematical model than human judgment. A human expert might be able to come up with a judgment in the form "it will rain," or "the weather will be mostly hot and sunny," but accurate predictions without employing a mathematical model are impossible.

Regarding qualitative forecasting, Tetlock noticed that while most experts' predictions were off, there was a small minority that consistently overperformed. Tetlock split people into the "foxes" and the "hedgehogs." These terms are based on the essay by Isaiah Berlin that references the ancient Greek poet Archilochus: "A fox knows many things, but a hedgehog one important thing." Foxes, in Tetlock's terminology, are able to gather information from multiple different sources and use more complex ways of thinking. Hedgehogs are monolithic, sticking to an opinion and trying to justify it, no matter what.

Tetlock later went on to start a new project called "The Good Judgment Project."[25] The goal of the project was to see whether average people could do better at predicting global events, compared to intelligence officers with classified information, and to also understand why some people are better at making predictions than others.

---

[25] www.gjopen.com/

Tetlock identified the best forecasters, called *superforecasters*, and wrote a book about this subject. According to Tetlock, these are the defining characteristics of superforecasters:[26]

1) Are actively open-minded, intellectually curious, enjoy puzzles, and like to think.

2) Are good with numbers. Even while numbers were not used in most forecasts of the study, superforecasters all have a facility with numbers.

3) Are good probabilistic thinkers—comfortable with giving answers in percentages. They like giving precise answers, but they recognize that there is uncertainty involved in every prediction.

4) Are non-religious. Superforecasters do not believe in supernatural forces or fate controlling the future.

5) Have less ego. Superforecasters are more likely to adjust beliefs on new information and have a "growth mindset." They always try to improve their skills through study. They also show self-discipline and are hard-working.

6) Consult multiple views and sources and regularly update forecasts based on new information.

But all this concerns qualitative forecasting. What can we learn from superforecasters about *quantitative* forecasting? Quantitative forecasting can seem esoteric to many people, since the methods applied are based on advanced mathematical concepts. These methods can extract pieces of information not detectable by humans. However, this doesn't make them immune to some of the fallacies and issues faced by qualitative forecasting.

One of the main issues faced by forecasting models is concept drift, discussed earlier. The modeler needs to exercise careful judgment and examine various sources of information in order to determine whether the system has experienced concept drift, and whether the data is of high enough quality.

---

[26] P. Tetlock, D. Gardner (2016), *Superforecasting: The Art and Science of Prediction*

Secondly, quantitative forecasting is often combined with qualitative forecasting. The predictions of the model can be combined with other factors, which a person will take into account when making decisions. Using both qualitative and quantitative methods can be a great way to make decisions.

Thirdly, an important part of quantitative forecasting is knowing when it works and when it doesn't. A wise general knows how to choose his battles, and a wise forecaster does the same. Recent advances in machine learning and technology have moved prediction into more exotic realms. This is exemplified by high-frequency trading. In high-frequency trading, the objective is to do multiple trades per second. While the exact movements of stocks over a longer time window might not be possible, predictions of the movements over sub-second periods could be. A similar approach can be followed by sports forecasters. If your goal is profit, there are countless odds markets to choose from. You can automate the search of those markets in which the algorithm seems to be more profitable.

The power of machine learning to extract every tiny bit of information from data, even from features that humans don't understand, has made it the standard in financial forecasting. However, there is no free lunch. The difficult part in machine learning for forecasting is testing. Because machine learning is so powerful, it tends to see patterns where none exist, much like humans do. That is why a sound testing strategy is of paramount importance. It is not uncommon for machine learning algorithms to display very bad performances in the real world. One of the most common issues is the lack of correct testing.

Forecasting, especially in finance, might seem like something relevant only to particular professions, and the average person might not find it a very fascinating subject. However, it doesn't take anything else other than a speculative bubble to see how everyone can very quickly turn into self-proclaimed experts and lose fortunes based on forecasts as to where the market will go. Recent history has given us some great examples.

In 2019, and then in 2022, the world was overtaken by a cryptocurrency mania. At the peak of this craze, Bitcoin prices soared up to $20k and then close to $50k.[27] Similarly, Ethereum went up to around $5k.[28]

---

[27] Retrieved from `www.coindesk.com/price/bitcoin` on March 8, 2019.
[28] Retrieved from `ethereumprice.org/` on March 8, 2019.

During the cryptocurrency craze, there were all sorts of predictions going on about how high the price would reach. There was, and still is, many people making all sorts of predictions, such as "bitcoin will reach $100k." A quick Google search will uncover plenty of articles.

In the end, the cryptocurrency craze ended up like the tulip mania, with all cryptocurrencies crashing and many people losing their money. So, what went wrong here? Based on what we've seen so far, how could have someone done better in this market?

After the crash, people came up with all sorts of explanations. Maybe the technology is not ready, or that it was a typical bubble, or that bitcoin prices will rise again. It's easy to come up with explanations after the fact. What is more challenging is coming up with correct forecasts before something happens.

There were two ways to beat the market. A qualitative forecaster would have to act like Tetlock's foxes. There were many signs that the cryptocurrency bull run was in fact a bubble:

1) The news being filled with statements from random individuals, showing up on non-reputable websites, making claims about how the prices of all the major cryptocurrencies will climb even higher.

2) The fact that there were no real collateral assets. Bitcoins do not represent anything tangible in the real world. It was just expensive because people agreed that they were willing to pay that price for it, and speculators thought they would be able to sell it to someone for more money (the "greater fool" theory).

3) You start hearing people who have nothing to do with cryptocurrencies or investment suddenly start talking about cryptocurrencies as if they are experts.

In this sense, the future of cryptocurrency prices could be predicted from past examples of other bubbles. The future was like the past, in a sense, but not as many people expected. Many investors believed that the future would be like yesterday's past of 10 percent growth, which was like yesterday yesterday's past of 8 percent growth, and so on. A fox would have been able to escape the herd mentality and see the bigger picture.

But if you know it's a bubble, you might still be willing to invest, capitalizing on your knowledge. Then again, how could you predict the exact point when the bubble will burst and get out before it happens? This might be impossible. Machine learning and forecasting models may have been able to help you in this front at first—on Google search at the time of writing returns more than two million results for the terms "cryptocurrency prediction machine learning"—but don't forget what we discussed earlier. It is better to employ machine learning in those situations where you know it will perform best. Predictions over small time intervals might be more successful than predicting where the price will be in two days' time.

Many of the issues with forecasting start at the human level, with the forecaster falling victim to fallacies and biases. We already discussed the issue of concept drift, and how the future might stop being like the past, because the underlying system has changed. Another one of the most common fallacies is the "positives examples fallacy," as I like to call it. Let's say that you are trying to figure out what successful companies or successful entrepreneurs have in common. How would you go about studying this? You can simply take a sample of successful companies or people and see what they have in common. This is what many people would do. However, in doing that, you have performed the positives examples fallacy.

When an algorithm tries to learn how to differentiate between examples, it needs variability. If all the companies you are investigating are successful, you can't tell which features distinguish the successful from the unsuccessful. You might be just confirming your biases. There might be many more companies, exactly the same as the ones you are investigating, that might have vanished for reasons of pure chance. Without comparison, it is simply impossible to know.

The truth is that while forecasting is difficult, it will always be needed and there will always be those who do it better than others. The Warren Buffets of the world know how to keep their calm, in spite of the temporary fluctuations of the market. The brightest quants will know how to extract every little piece of information from the stock market and will update their models the second those patterns stop working.

Forecasting is an extremely powerful and useful application, but to most people it can seem elusive. Even the best forecasters in the world can't be 100 percent right. This can make one ponder the limits of our predictions. We will ever be able to know everything, or at the very least, can we know what we know and what we don't know?

CHAPTER 11

The Limits of Prediction (Part A): A Futile Pursuit?

"Prediction is very difficult, especially if it's about the future."

—Nils Bohr, Nobel laureate in Physics, 1922

So far we've seen some of the major approaches in data science, and we discussed many of the challenges that we face when trying to predict the future or make algorithms that learn from data. It's natural for someone to ask, what exactly *can* we predict and what can't we? How can we have faith in our predictions? What are the limits of our knowledge? Is it always the case that we won't know what we don't know, or can we, at least, get an understanding of how much nature allows us to learn about the future?

There are different ways to approach how to solve these questions. This chapter talks about two. One is a top-down, abstractive approach, the other is a bottom-up reductionist approach.

The *top-down, abstractive approach* is the framework of learning theory in machine learning. Learning theory concerns itself exactly with this type of question—"How much can a learning algorithm learn, and how well can it perform?" But instead of studying this through experimentation (which is the main approach in machine learning), it studies through theorems that offer guarantees around the performance of models.

In essence, "learning theory" is the study of the uncertainty of the models that we build to deal with uncertainty.

This is a pretty ambitious goal. You don't know anything about the data, the problem, or the algorithm, but you do know something about how well the model will perform on average.

© Stylianos Kampakis 2023
S. Kampakis, *Predicting the Unknown*, https://doi.org/10.1007/978-1-4842-9505-2_11

Another approach is to ignore theory altogether and focus instead on creating a reduced, simplified version of the full system that we're studying (the "reductionist" approach). This is the core tenet of modeling and the Monte Carlo simulations. This approach is quite different from statistical modeling and machine learning, as it focuses on simulating aspects of the world, rather than just abstracting most of it through constructs like probability distributions. This chapter talks about modeling physical systems, as well as social systems, that are modeled through agent-based modeling. You'll see how modeling is a great way to control both aleatoric and epistemic uncertainty, but I'll also challenge some of the shortcomings of this approach, and the limitations that nature places on our capacity for knowledge.

Learning Theory: What Can We Know About What We Don't Know?

Learning theory tries to answer, in a mathematical way, questions about how much an algorithm can learn and how well it can perform. Even to the initiated data scientist, machine learning can look a bit like magic. You feed data into a black box, and it comes back with predictions. It's not always clear which algorithm performs best and *why*. Learning theory is an attempt to answer those questions.

Learning theory is similar to some of the theorems we saw earlier about probabilities, like Chebyshev's inequality. The fact that we can discover universal laws of mathematics that hold true under all circumstances is a testament to the power of human reasoning. Learning theory is essentially an attempt to do something similar for machines. The questions concerned are as follows:

1) Which are the best algorithms for any given problem?

2) How much data do we need?

3) What is the best performance we can expect?

4) Why does Occam's razor work?

If you notice, these are similar to the kind of questions that forecasters tried to answer during the M-competitions. Learning theory is the theoretical counterpart of the competition approach.

There are different approaches in learning theory, but the most popular one is the PAC/VC theory. PAC stands for Probably Approximately Correct and is a theory developed by Leslie Valiant[1] in 1984. Remember the problem of induction? After seeing 10,000 white swans, but never a black swan, you've formed a hypothesis that all swans are white; but how do you know that you are absolutely right?

Let's go back to the example you saw in the first chapter. You time-travel to the past, and you're trying to discern which foods are edible. You come up with a hypothesis that foods with certain properties (e.g., foods that are red) are edible. The hypothesis might be correct only a certain percentage of the time (e.g., only 90 percent of the red fruits are edible). This is the approximate part. The theory is not totally correct, but it's approximately correct.

So where does the "probably" part come from? Besides this theory, you might come up with other theories about edible foods. You might even come up with theories about edible liquids, besides water (e.g., is orange juice edible or not?). If most of the theories you come up with are approximately correct, then you are a "probably approximate correct learner." This means that most of the time, most of the theories are good enough.

Replace yourself in this example with an AI algorithm, and you can see how this directly extends to machine learning. We want our machine learning algorithms to be correct as often as possible. We want them to be probably correct, most of the time. That is probably approximately correct.

Valiant was awarded a Turing Award[2] for his contribution to computer science. This is what he had to say about how he came up with his theory:[3]

> "At the time I started working on it [in the 1980s], people were already investigating machine learning, but there was no consensus on what kind of thing 'learning' was. [...] So I thought that learning should have some sort of theory. [...] Learning must be something statistical, but it's also something computational. I needed some theory which combined both computation and statistics to explain what the phenomenon was."

I mentioned, however, that the current framework is called PAC/VC. VC stands for Vapnik-Chervonenkis. Vladimir Vapnik and Alexey Chervonenkis came up with the VC theory. The VC theory is a way to express the power of a machine learning

---

[1] The original paper, for those interested, is located at: `web.mit.edu/6.435/www/Valiant84.pdf`
[2] `amturing.acm.org/award_winners/valiant_2612174.cfm`
[3] `www.quantamagazine.org/the-hidden-algorithms-underlying-life-20160128/`

algorithm. Given something called the VC dimension (which is specific to each machine learning algorithm), and the performance on the model on some data, we can get some guarantees around the performance of the model in the real world.

The VC framework, much like the PAC framework, delves into difficult mathematical formulations to prove their points. However, for the purposes of the reader who is interested in understanding how these theories help us fight uncertainty (without deep-diving into esoteric mathematical formula), the main take-away is that we have mathematical tools that in an abstract way can tell us how well we can model uncertainty. That is, we might face uncertainty around our own models of uncertainty, but this can be controlled to some extent.

However, those theories have not found much use in the real world. Whereas the VC framework gave rise to a popular algorithm called "support vector machine," in practice the formulas are not easy to use for a variety of reasons. Plus, support vector machines have fallen out of fashion in favor of more complicated models, like deep learning.

When Vladimir Vapnik came up with his theories between 1992-1993,[4] neural networks were still popular. However, one of their main drawbacks was that they acted as black boxes. No one really knew exactly *why* they were working. Vladimir Vapnik's support vector machines were better founded theoretically, and they were very transparent as to how and why they work. This led to a huge wave of research on support vector machines and related algorithms. However, those algorithms suffered from some basic problems.

First, it took a long time to run them. They were computationally intensive. Secondly, the researcher had to spend lots of time optimizing the hyperparameters. The hyperparameters are like the tuning knobs of algorithms. Finding the right settings can be a long process. This meant that, in practice, they weren't always a good choice.

Quite ironically, deep learning did the exact opposite. It works well, but with no real theory backing it up. However, deep learning has its own challenges too. It's a very popular and powerful set of models, but no one really knows exactly why they're supposed to work. The success of deep learning algorithms has inspired a whole new field of research around them. For example, some researchers try to use tools from statistical mechanics to explain why deep neural networks work so well.[5] Others, like Yoshua Bengio—one of the fathers of deep learning—proposed that it might be due to the way that they are trained.[6] The truth is, we still don't have a conclusive answer. However, we might find one in the next few years.

---

[4] The first version came out in 1963, but it was not until 30 years later that they became usable.
[5] arxiv.org/pdf/1710.09553.pdf
[6] arxiv.org/abs/1710.05468

In any case, learning theory has given us lots of useful tools and methods, which has helped improve machine learning algorithms. Maybe the quest to be able to tame the uncertainty of our models completely is a futile one, but in the process we can learn a useful thing or two.

Something worth noting about the PAC framework is that Valiant extended this to the theory of evolution. According to Valiant, the biological world is computational at its core. In his 2013 book called *Probably Approximately Correct,*[7] he expanded the concept of an algorithm to that of an "ecorithm."

An ecorithm is basically an algorithm, in that it receives feedback from an unpredictable and complex world, like the one we encounter in real life. So, each species in life needs to adapt to its environment, by maximizing the probability of its survival and procreation. The adaptation happens through the process of evolution, which is essentially a learning process. Evolution receives feedback from the environment, and then adapts the species in a way that improves the species' probability of survival. According to Valiant, evolution belongs to a class of algorithms called "statistical query" algorithms, and, since it is an evolutionary process, it can be analyzed through the PAC framework.

It's not easy to say whether this theory is correct or not, but it demonstrates how the study of uncertainty raises interesting questions about many other aspects of our lives and the world around us. In the first chapter, we discussed how someone would have to deal with extreme uncertainty in order to survive in the wilderness, while trying to determine what foods are edible and which ones are not. This simple question demonstrates the impact of uncertainty, the depth of questions around studying it, and how complicated the theories and formula that try to explain it can get.

Indeed, uncertainty exists the moment an intelligent creature decides to make a decision for which the outcome is not certain, and so permeates many of the actions that intelligent life has to take. However, what about physical systems? The study of physical systems was connected, for a long time, to mechanics and concepts like the equation of gravity. But uncertainty can exist everywhere, including the physical world, and this is the topic of the next section.

---

[7] www.probablyapproximatelycorrect.com/

Monte Carlo Simulations: What Does a Casino Have to Do with Science?

"In the casino, the cardinal rule is to keep them playing and to keep them coming back. The longer they play, the more they lose, and in the end, we get them all."

—Robert De Niro in *Casino*

Casinos and data science have a long history together. Many examples in introductory probability theory textbooks are phrased in terms of games of chance. Besides the fact that games of chance provide an idealized version of probability and are relatively easy to understand, many of the early advances in probability theory came from trying to solve games of chance.

Gerolamo Cardano (1501–1576) used probability to analyze games of chance. He wrote the book *Liber de Ludo Aleae* (*Book on Games of Chance*), which used probability theory to study games of chance. This is considered the first systematic study of probability.

Pierre de Fermat (1607–1665), in collaboration with Blaise Pascal (1623–1662), created the foundation of the modern theory of probability. The collaboration came as part of an effort to help the French nobleman Antoine Gombaud understand why betting on rolling at least one six in four throws of a die is not the same as rolling one double-six in 24 throws of two dice. In fact, he was winning money on the first bet, but losing money on the second. A happy by-product of Gombaud's wealth-fueled hubris was also the foundation of modern calculus.

Random games have played a huge role in the development of probability theory, and Monte Carlo can be seen as part of this tradition. The term "Monte Carlo" is of course inspired from the famous casino region in Monaco, which has been operating as a gambling hot-spot since the mid 1800s. The Monte Carlo approach was devised by Stanislaw Ulam (1909–1984), while working in nuclear weapons at Los Alamos. Physicists at that time had issues around using deterministic methods to come up with solutions for various problems. Ulam came up with the idea to simply run simulations to find the solution. Since the people at Los Alamos were working on a top-secret project, they came up with the name "Monte Carlo method" in order to hide their methodology.

The creation of the computer in the late 1940s unlocked the possibility of running simulations to understand physical processes for the first time. Monte Carlo simulations played a role in the creation of both the atomic and the hydrogen bomb. As computational power increased, this method became more popular in various numerical problems, like physical simulations. Since the turn of the millennium, it has also been used extensively in order to solve the problem of Bayesian inference, which we looked at earlier.

The core concept of Monte Carlo simulation is deceptively simple. You take a model of the world, and parts of this model include some random components. These represent aleatoric uncertainty about our system. We've modeled these using probability distributions. We'd like to find an exact solution, but it's proving very difficult. So, instead of doing that, we run a large number of simulations and see what we come up with. In other words, the more you try, the more you learn.

Each simulation is a different scenario that could happen in the real world. A large number of simulations can give us an idea of:

1) What the average scenario looks like.

2) What the variance among different scenarios is.

It's a simple but powerful idea. It might sound intuitive these days, especially as we have access to personal computers, but in the 1940s when it was first devised, it was a revolutionary concept. These days, Monte Carlo simulations are used in many fields, from fluid dynamics to finance.

It's worth noting that statistical approaches to science had already been devised many years before the discovery of Monte Carlo simulations. It's just that similar ideas were not necessarily widespread in all scientific areas. Statistical mechanics is one of the most important parts of modern physics. The core concept behind statistical mechanics is that systems with many interacting parts are difficult to describe through exact equations. Therefore, statistical approaches must be devised.

Statistical mechanics aims to understand the macroscopic behavior of systems, through the properties of their microscopic constituents. A good example is a mass of gas. Let's say that we're interested in the macroscopic properties of this mass. That is, we care about things like the density or the pressure. We don't care about modeling each individual molecule. We can, however, use the laws of probability to model the average behavior of the molecules.

This is a great concept, because it lets us take out the unnecessary details to physical objects and focus instead on the things that we care about. Think about the computational complexity of modeling each individual molecule—getting an average of the behavior of the molecules is clearly a far easier approach. Probability theory is one of the best ways we have to quantify uncertainty, and statistical mechanics is another success. Statistical mechanics are applied to a large range of physical phenomena, from electric conductivity, to fluids and solids.

The foundations of statistical mechanics were set by James Clerk Maxwell (1831–1879) and Ludwig E. Boltzmann (1844–1906). Maxwell was a Scottish scientist who made huge contributions to physics. His greatest contribution has been in the theory of electromagnetism. Maxwell is responsible for discovering that electric and magnetic fields move like waves at the speed of light, which made possible the discovery of radio waves. Einstein described Maxwell's work as the "most profound and the most fruitful that physics has experienced since the time of Newton."

Something that is less well known is that Maxwell also contributed to control theory, which can be seen as a precursor to artificial intelligence. Control theory deals with the control of mechanical systems, and Maxwell, in 1868, presented a formal analysis of a centrifugal governor. This is a device that controls the speed of an engine through the amount of fuel that is admitted. While this is far from being AI, it is a system that responds to input (the speed of an engine), by a corresponding output (the quantity of fuel), so the overall concept it is not too dissimilar from a crude reinforcement learning algorithm, for example.

Ludwig Boltzmann (1844–1906) was an Austrian physicist and philosopher. Boltzmann is famous for his multiple contributions to physics, which also laid the foundation of statistical mechanics. Boltzmann's most famous equation is the entropy formula $S = k \log W$, which is also written on his tombstone and describes the entropy S of an ideal gas system, based on W, the total number of ways in which this system can be organized, and k, which is the Boltzmann constant (but we don't have to concern ourselves too much with that here).

Boltzmann at his time was a firm proponent of the theory that molecules are composed of atoms. However, this theory was largely unpopular at that time. In the end he was actually one of the few scientists that supported it. Boltzmann was the only person who actually knew the truth about how nature works, but his truth would not be accepted. He was still a popular scientist, but this must have left a mark on him. The failure to prove his theories, in combination with bipolar disorder, led him to hang himself in 1906. Sadly, atoms were proven incontrovertibly just a few years later.

Boltzmann's work also relates to the famous problem of the thermodynamic asymmetry of time—that is, time seems to flow only in one direction. A glass can break, but we can't reverse time and make it whole again. The heat dissipates from a hot cup of coffee, and it never goes back into a coffee, unless we expend energy to warm it up. The second law of thermodynamics dictates that entropy of an isolated system can increase, but not decrease, and this creates the arrow of time.

While his work is not directly related to machine learning, Boltzmann lent his name to a machine learning model called the "Boltzmann machine," which was inspired by his work in thermodynamics.

The actual term "statistical mechanics" was coined Josiah Willard Gibbs (1839–1903), another one of the most prominent physicists of all time. His name is also related to "Gibbs sampling," one of the most successful Monte Carlo methods. While he was not the one to describe the method (the method was described by the mathematician brothers Stuart and Donald Geman in 1984), the concept behind this method was based on Gibbs' work.

There is an interesting observation to be made here. Statistical mechanics became a very successful theory, specifically because physicists moved from modeling the exact equations to using probabilities. Secondly, Monte Carlo simulations provided a great way to include uncertainty into our models.

The methods developed by statistical mechanics reshaped the face of science, with countless applications in chemistry and engineering. Concepts from statistical mechanics have also been used by machine learning, for example in information entropy and neural networks. Monte Carlo simulations are also used in many different fields, from investment to meteorology, and are the basic inference mechanism of Bayesian models.

Simulation opens up all kinds of possibilities outside of physical system and poses some complex and ethically divisive questions. If we had so much success with simulations of such systems, why not also try simulations of social systems? Social systems are ridden with uncertainty. Humans can be quite unpredictable, and societies can seem, sometimes, even more so. Why not use a simulation to study this complexity? What might we learn if we did? And what could possibly go wrong?

CHAPTER 12

The Limits of Prediction (Part B): Game Theory, Agent-based Modeling and Complexity (Actions and Reactions)

"It is just as foolish to complain that people are selfish and treacherous as it is to complain that the magnetic field does not increase unless the electric field has a curl. Both are laws of nature."

—John von Neumann

Arguably, there is no other mathematical theory that has done as much to destroy the illusion of equilibrium in social dynamics as game theory. Game theory was originally devised by John von Neumann and Oskar Morgenstern in 1944.[1] What game theory contributed to science was that it assumed that the entities involved in a system had agency and their actions would affect the outcome of the game. In other words, everyone

---

[1] J. von Neumann, O. Morgenstern, 1944, *Theory of Games and Economic Behavior*

© Stylianos Kampakis 2023
S. Kampakis, *Predicting the Unknown*, https://doi.org/10.1007/978-1-4842-9505-2_12

has an agenda. Each player has a set of goals but also needs to figure out what the other player will do. Up until that point, uncertainty was something that was only part of nature. Game theory demonstrated that the intentions and strategies of humans are also a cause of uncertainty and, in many social systems, humans are actually the prime source of uncertainty.

The most famous example of a game theory is the prisoner's dilemma, originally framed by Merrill Flood and Melvin Dresher in 1950. The prisoner's dilemma represents a fictional case where two culprits are being interrogated by the police. They are faced with the following options: to confess or not confess. The payouts for each player depend on the actions of both players. So, the following outcomes are possible:[2]

1. Neither player confesses: they both get one year in jail.

2. One player confesses, the other does not: The player who confessed goes free (which is a reward for cooperating with the authorities), and the other player gets three years in jail.

3. Both players confess: They both get five years in jail.

This game is very tricky. The best outcome for an individual player is for this player to confess, and the other to not confess. The best outcome from a moral perspective would be that neither confesses. However, the final outcome, assuming the players are rational, is that both players confess and end up in the worst possible scenario.

Game theory is used in the analysis of oligopolies, which are markets where there are only a few companies competing—think oligarchs combined with monopolies. For example, a price war is a typical example of prisoner's dilemma applied in real life. Game theory can be applied in many other situations, from understanding how interest rates should be taxed, to nuclear disarmament. The latter is another real-world example of the prisoner's dilemma. Whereas it might be safer for all countries to destroy all their nuclear weapons, if an individual country doesn't oblige, then it will automatically become a superpower.

Game theory has had a big impact on economics and mathematics. This impact can be seen by the fact that it has also been the subject of 11 Nobel prizes. One of the prize winners was John Nash, whose story was popularized in the movie *A Beautiful Mind* staring Russell Crow.

---

[2] The payouts in this example are arbitrary. What is important is the structure of the payouts (e.g., both players confessing is worse than not confessing).

While game theory demonstrated the complexity of social interactions and how they
can be a source of uncertainty, it also demonstrated how, even in cases of uncertainty,
we can be certain of the outcome, as certain games can achieve equilibrium states[3]
(given some assumptions). This approach to studying uncertainty is directly related to
agent-based modeling, to which I'll come back shortly.

The convenient assumption is that the agents will behave rationally, which allows
us to make predictions about the equilibrium states of various outcomes. However, as
Kahneman and Tversky (you will read about them shortly) discovered, humans are
anything but rational—at least in the game theoretic sense. Real behavior is even more
complicated than game theory could predict, which is why complexity theorists (who
you'll also visit shortly) used game theory.

Agent-based Modeling: Crafting Artificial Worlds

Agent-based modeling is a bit like trying to create a video game of the world—*The
Sims*, but with theory. In agent-based modeling, the goal is not so much to abstract
relationships as it is to create an accurate representation of the world *in silico*—that
is, inside a computer. Some of the first agent-based models can be attributed to von
Neumann's cellular automata (I'll talk about these in a later section), but probably the
first realistic agent-based model was Schelling's segregation model.

Thomas Schelling (1921–2016) was an American economist and professor of foreign
policy. While he was at Harvard he became interested in how segregation emerges. So,
he created the following simple model. There are agents that live in a world made up
of square lattices, much like a board game. The agents are very simple entities, which
represent abstractions of humans. The agents are split into two different colors, let's
say red and blue. The game proceeds in rounds, and in each round each agent gauges
whether it is happy or not. For an agent to be happy, a percentage of its neighbors
need to be of the same color. If not, then the agent will attempt to move to some other
empty cell.

---

[3] An equilibrium state is a state of balance. Once a game reaches the equilibrium state, no further
change take place.

What we can see in this game is that even when the agents are very tolerable, the game will converge to a state where there are clear neighborhoods consisting of agents of the same color. You can see this in Figure 12-1. In this particular instantiation of the game, the agents are happy as long as 40 percent of their neighbors are of the same color. But even this forces the creation of segregated neighborhoods.

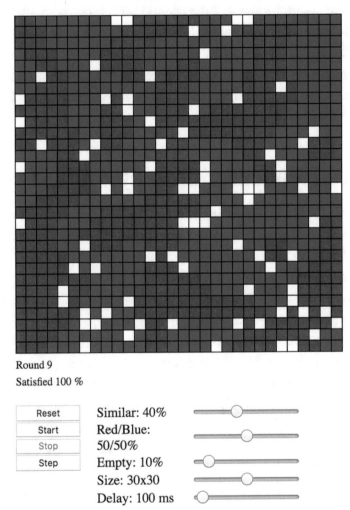

Round 9
Satisfied 100 %

Reset	Similar: 40%	——————O——————
Start	Red/Blue:	——————O——————
Stop	50/50%	
Step	Empty: 10%	—O—————————
	Size: 30x30	—————O———
	Delay: 100 ms	O———————

Figure 12-1. *Schelling's segregation model*[4]

[4] Screenshot taken from this implementation: nifty.stanford.edu/2014/mccown-schelling-model-segregation/

This is particularly interesting because it demonstrates that segregation is not necessarily the result of prejudice on the part of the individuals. Even when individuals do not even care that much about the majority of their neighbors being like them, segregation can still take place. The behavior of Schelling's model is characteristic of complex systems that can display emergent behavior. I talk about what this is shortly, but the main idea is that the whole is more than just the sum of its parts.

Another famous agent-based model is Sugarscape. Sugarscape was created by Joshua M. Epstein, who is a Professor of Epidemiology at the New York University College of Global Public Health, and Robert Axtell, who is a professor of computational social science at George Mason University. Sugarscape is an artificial world that consists of a 51x51 square lattice. The agents have various rules to govern their interactions, which are based on simplified versions of our own world. The main resource in this world is "sugar," which the agents have to consume. The agents can also reproduce and die, and in some more complicated versions of the world, even barter.

Epstein and Axtell published their results in a book called *Growing Artificial Societies: Social Science from the Bottom Up*, which was first published in 1994. The main point of the model was to demonstrate how complex dynamics can emerge from simple rules, and how those dynamics and rules can be very similar to the dynamics and rules we observe in the real world.[5]

Research on agent-based modeling moved on further and was applied to various problems. In 1999, Nigel Gilbert published the first textbook in the field, called *Social Simulation: Simulation for the Social Scientist*, and he also established the first journal for agent-based modeling in the social sciences, the *Journal of Artificial Societies and Social Simulation*. A search on Google Scholar for "agent-based model" returns over 4.5 million results, for models relating to anything from supply chain, to military operations. The Santa Fe Institute,[6] which was established in 1984, is also another prominent name in this area, with multiple contributions to the field, especially the field of agent-based models in economics.

---

[5] Epstein, J. M., & Axtell, R. (1996). *Growing Artificial Societies: Social Science from the Bottom Up.* Brookings Institution Press.

[6] www.santafe.edu/

Computer simulations have played a prominent role in the physical sciences, and agent-based models are trying to do the same for social sciences. Agent-based models have a unique property in that we don't have to come up with mathematical models or abstract theories as to how the world works. We can just take small pieces of theories, compile them, and then let them run in a simulation. Understanding how human psychology affects economic choices is a fascinating and complex problem, and the process of answering it gave rise to the field of behavioral economics.

Understanding how the findings of behavioral economics affect the wider context of an economy is a much more difficult problem. However, using an agent-based model, we can create agents that simulate real humans, using any and all theories of psychology available to us, and then see how their interactions can generate all kinds of phenomena.

Agent-based modeling can be very powerful. Many complex systems in nature—like social systems—are characterized by feedback loops and dynamics, which make them difficult to control or predict. In fact, the case for many natural systems is that they do not even have to follow complicated rules to display this kind of complex behavior. In this kind of situation, simpler deterministic or analytical models are not as useful. Agent-based modeling is one way to accept the limits of our methods and knowledge, but also come up with a new way to attack the same problems. It is a clearly pragmatic approach, in direct contrast to computational learning theory, which I talked about earlier.

While agent-based modeling (as well as these observations) had been around since the 1940s, it was during the 1990s when this approach started taking the form of a theory. This effort was termed "complexity theory."

Complexity Theory: Simulation vs the Limits of Prediction

"Bottomless wonders spring from simple rules, which are repeated without end."

—Benoit Mandelbrot, 20th century mathematician and polymath

The popularity of complexity theory boomed during the 1990s and early 2000s. Complexity theory exclusively uses simulations and was one of those scientific theories that tried to explain everything, but in the end, well... it didn't. However, in the process, it gave us a good understanding of our limitations in studying uncertainty in many important phenomena, social or physical.

As the name implies, complexity theory is the study of complex systems. What are complex systems exactly? There is no clear definition, but we can describe them. Complex systems are systems that are not easy to analyze through traditional analytical techniques. Complex systems have many interacting components, and they can behave in unpredictable ways. These two points are key. Analytical equations, like the ones used, for example, to describe simple physical phenomena, cannot be used to understand what the next state of the system will be. This means that a set of new tools had to be devised to try and do that.

Why the fuss about complex systems? Some of the most interesting (and important) systems we encounter are complex: the economy, social systems of all kinds, even some systems we encounter in nature like neural networks and ant colonies. Understanding how these evolve and adapt would help us have a greater control over the economy, unlock the mysteries of the brain, and even eliminate global conflict. Or at least, this is how the theory goes.

One of the most popular concepts in complexity theory is self-organized criticality. What is self-organized criticality? I explain this with the sand pile example, probably the most famous example in complexity theory. Imagine a pile of sand where we keep adding sand, one grain at a time. There is a certain point where the next grain of sand we add will cause an avalanche. Many grains of sand will roll down the pile, restructuring it, all as a result of an individual grain of sand that was added.

When the pile of sand is at a point where adding one more grain will cause an avalanche, we say that it is at the point of self-organized criticality. The pile is in equilibrium before we drop that single grain of sand. However, a miniscule change will completely reorganize it. But after the pile is reorganized, after the avalanche, it will reach a new equilibrium state. It will, once again, take a considerable number of grains of sand before a new avalanche is created.

We talked earlier about the butterfly effect—how a butterfly flapping its wings in Brazil can cause tornados in Texas. This example was used in chaos theory to demonstrate the interconnectivity of the world. However, this interconnectivity takes a different meaning in complexity theory. The example of the pile of sand is not trying to demonstrate the chaotic nature of the world. Rather, the theory behind self-organized criticality is that systems that are at the edge of criticality can rapidly change their structure, reaching a new equilibrium state. However, the exact details of this (when the transition takes place, and what the system looks like after the transition) are not easily predictable.

The concept of self-organized criticality was first noted by Per Bak, Chao Tang, and Kurt Wiesenfeld in 1987.[7] They observed that various systems in nature, including cellular automata which I talk about next, have some common properties like fractal structure, pink noise, and power laws. Fractals are mathematical objects that have a fractional dimension. For example, instead of existing in one, two, or three dimensions, they exist in 2.43 dimensions. The major characteristic of fractals is that they are self-similar. No matter how much you zoom in or out on a fractal, the observed patterns are similar. In an analogy with society, the economic organization within a household might be similar to that of a borough, which is similar to that of a city, a larger county, a country, or an international organization like the EU. See Figure 12-2.

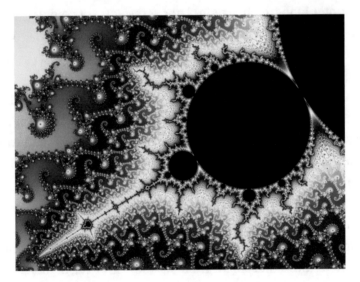

Figure 12-2. *Example of a fractal, the famous Mandelbrot set*

The first person to make the connection between fractal geometry, economics, and finance was Benoit Mandelbrot. In his book, *The Misbehavior of Markets: A Fractal View of Financial Turbulence,* he explained how market fluctuations can be better understood through fractal geometry than existing economic theories. His work was also a precursor to complexity theory, as he had demonstrated how complex patterns can emerge from simple rules.

---

[7] Bak, P., Tang, C. and Wiesenfeld, K. (1987). "Self-organized criticality: an explanation of 1/f noise." *Physical Review Letters.* 59 (4): 381–384.

The sand pile analogy might look somewhat unrelated to more meaningful systems. However, this concept of self-organized criticality shows up in many real systems. The economy is one of the all-time favorites of complexity theory, to the extent that "complexity economics" was even used as a term to describe the application of those theories in economics.

In a paper in 1992, Bak described[8] how an economy can be seen as a dynamic system, with interacting feedback loops. Whereas in classical economics small deviations from the mean will eventually converge back, in dynamic systems small perturbations can combine in myriad ways. One way is that they can cancel each other out and dissipate (which explains the behavior expected by mainstream economists), but they can also combine in positive feedback loops, which can eventually cause financial crises or other major cataclysmic phenomena. Economies are at a state of self-organized criticality that can help them adapt to changes, but this can also occasionally cause major disturbances.

These views come in direct conflict to orthodox economics, which assume that the economy exists in an equilibrium, and crises are rare. Classical economics is an equilibrium approach. Classical economists seek strategies and methods that can provide a better equilibrium. Complexity economists, conversely, study the feedback of dynamic loops that take place as a result of the interactions among agents. It is a non-equilibrium approach, and it represents a complete paradigm shift compared to previous approaches. In Mandelbrot's words:

> "Most economists, when modeling market behavior, tend to
> sweep major fluctuations under the rug and assume they are
> anomalies. What I have found is that major rises and falls in prices
> are actually inevitable."[9]

It has been argued that self-organized criticality is also evident in biological neural networks. The point of self-organized criticality is a balance between order and disorder. This balance assists adaptation and dissemination of information. While not everyone accepts this theory, there is evidence that this could indeed be the case.[10]

---

[8] Bak, P., Chen, K., Scheinkman, J. A., & Woodford, M. (1992, May). "Self-Organized Criticality and Fluctuations in Economics." Santa Fe, NM: Santa Fe Institute.

[9] foreignpolicy.com/2009/06/21/benoit-mandelbrot/

[10] For a good summary in plain language, read www.quantamagazine.org/toward-a-theory-of-self-organized-criticality-in-the-brain-20140403/

Another example of a complex self-organized system is an ant colony. While ants are very simple creatures, their colonies are highly complicated societies that can adapt to many different circumstances and respond to stimuli in the environment as one body. It's been speculated that ant colonies exist in a state of self-organized criticality.

Another key term behind complexity theory is "emergence." Emergence exists when the total is more than the sum of its parts. Complex systems can interact in ways and generate phenomena that are completely different from their constituent parts.

The most famous example of this is consciousness. The problem of consciousness has been described as one of the most challenging and critical problems of modern science. Even though some people might disagree with this statement, I would say that the study of consciousness is somewhat like the scientific study of what religions have called "the soul."

What exactly is consciousness? Consciousness describes all subjective feelings that a being can experience. There are different kinds of consciousness. The simplest one is simply having sensations. We can't know, for example, what happens inside an insect's mind. However, we know that insects have brains like us, and we can be fairly certain that they must feel sensations like touch and temperature.

A higher kind of consciousness is self-awareness: knowing that you exist. Most animals do NOT possess this capability. It's mammals like primates and dolphins that have passed the "mirror-test."[11]

So, how does emergence relate to consciousness? If we are to study individual neurons, there is nothing to suggest that neurons connected in the same way can generate anything remotely related to consciousness. Let's take the reductionist view of science: Everything can be broken down into its constituent parts, and this is how we can understand the world.

---

[11] The mirror-test was devised by Gordon Gallup Jr. in 1970 (Gallup, G. G. (1970). Chimpanzees: self-recognition. *Science*, 167(3914), 86-87). An animal is placed in front of a mirror. At a later time, a mark will be made on an animal's body and then it will be placed in front of the mirror again. Animals that pass the mirror-test will observe the mark on the mirror and might even try to change their position of their body to get a better view. Compare this with the behavior of dogs, for example, which might aggressively bark at the mirror, thinking that it is another dog staring at them.

Neurons are composed of cellular bodies. They transmit information to each
other through synapses and axons and the use of electrochemical reactions. Neurons
themselves are composed of molecules and atoms. How is it that many connected
neurons together generate something that can have feelings and sensations, that is,
a human?

This question gets more complicated the more you think about it. You can assume
that other people are conscious, based on analogical reasoning. You have consciousness,
and these other creatures look like you, so they must be conscious too. Do other animals
have consciousness? Well, they do have a brain and a nervous system, so it is reasonable
to assume that they do, even if researchers in the past, like behaviorists, completely
ignored the role of the mind in many living organisms.

However, what about creatures that *don't* look like us, but still have neurons and
brains, albeit different to ours? Is a worm conscious? Are creatures without a brain,
but with a nervous system, conscious? Are individual neurons conscious? Maybe this
question is nonsensical, but how can you know? If many neurons together for some
reason produce consciousness, why can't each individual neuron, or small groups of
neurons, be conscious as well?

We can also perform the following thought experiment. Let's say that an evil wizard/
mad scientist was studying consciousness and was interested in answering these
exact questions. The evil wizard/mad scientist decides to do so by taking a human and
removing neurons, one at a time, and then using their special consciousness-measuring
device to see whether the human is still conscious or not. At how many neurons does a
human stop being conscious? Does the connectivity play a role?

We can also reverse the question. There have even been attempts to use biological
neurons for computation. William Ditto worked on this concept while he was at Georgia
Institute of Technology. This has been called a "wetware" computer. Ditto managed
to manufacture a computer from leech neurons that could perform addition.[12] So,
instead of having normal computers with CPUs made out of silicon, we could simply use
neurons cultivated *in vitro*. The question is, would this machine be conscious? How do
we know that a computer created in a lab that consists of 100 neurons does or doesn't
feel pain or joy or some kind of otherworldly experience to which we don't have access
to, just because it can't communicate with us?

---

[12] This idea was largely abandoned, but there is at least one company working on this concept at
the time of writing.

At this point, we start wading into a quagmire of uncanny and uncharted territory, which raises a plethora of philosophical, scientific, and ethical questions—not the least of which is, do we even exist at all? Or was Descartes right all along? And will the falling tree make a sound if no one is there to hear it?

David Chalmers, one of the most prominent philosophers of the mind, has broken down the problem of consciousness[13] into two sub-problems: the easy problem and the hard problem of consciousness. By the term "easy problems" or "soft problems," he refers to all those quandaries relating to consciousness that can be answered by the current methods of science. For example, problems like discriminating stimuli and reporting emotions or behaviors are soft problems. We can either record this information, or ask the subjects to report it.

The hard problem concerns the how and the why. How is consciousness generated in the brain, and why? When a creature perceives the world, there is "something that it is like" to be that creature. A bat is a bat because being a bat means perceiving the world through echolocation. This subjective aspect is experience. In his own words:[14]

> "The really hard problem of consciousness is the problem of experience."

> "It is widely agreed that experience arises from a physical basis, but we have no good explanation of why and how it so arises. Why should physical processing give rise to a rich inner life at all? It seems objectively unreasonable that it should, and yet it does. If any problem qualifies as the problem of consciousness, it is this one."

The easy problem is about finding the neural correlates of consciousness. That is, which parts of the brain generate what sensations? Through modern advances in methods such as functional magnetic resonance imaging, we can also find neural correlates of various events. For example, subjects can report they feel a particular emotion, and then we can identify this emotion in the brain. However, this doesn't answer which specific neurons or neuronal events can generate conscious experience, and why.

---

[13] Chalmers, D. J. (1995). "Facing Up to the Problem of Consciousness." *Journal of Consciousness Studies,* 2(3), 200-219.

[14] consc.net/papers/facing.html/

According to complexity theory, consciousness is an *emergent* phenomenon. That is, there is nothing that foretells that a set of neurons connected together should display the property of consciousness. It is a property that appears as the result of the interconnectivity of a number of neurons in specific ways.

Are there any other emergent phenomena? Yes, there are. The formation of social structures, like cultures and economies, are also considered emergent phenomena. The organization of various insects, like ant colonies and bees, are considered emergent phenomena. The common element between all of them is that the whole is more than just the sum of its parts.

Studying Complexity Is a Complex Endeavor

"The complexity of things—the things within things—just seems to be end-less. I mean nothing is easy, nothing is simple."

—Alice Munro, Nobel laureate in literature

At the peak of its popularity, complexity science researchers were enthusiastic about the potential of these new theories to explain phenomena in radically new and more informative ways. Given that many of the aforementioned problems were classified as emergent phenomena, it becomes easy to see why it was a very attractive thought to believe that we could explain a wide range of important scientific questions through a general theory of emergence.

However, this didn't happen. Why? Well, one issue with theories that try to explain everything is that quite often they end up explaining nothing. Complexity theorists came up with many interesting examples and formulas, but most of them had no predictive value of actual systems. The formulas behaved like real systems, but as we discussed earlier, a key component of complex systems is the fact that they are difficult to predict. Complexity theorists had success explaining the qualitative properties of such systems, but not much success predicting the quantitative properties.

One of the most prominent examples of how complexity science was, at the same time, both spectacular and a complete failure was Stephen Wolfram's *A New Kind of Science*. Stephen Wolfram was a child prodigy who finished his PhD at the California Institute of Technology at the age of 19. He quit academia in order to create the popular mathematics software Mathematica. His company, Wolfram Research, was then founded in 1987, and it eventually became pretty successful. Among other things, Wolfram

Research released a search engine called Wolfram Alpha,[15] which provides intelligent answers to math- and data-related problems. You can use it to run some of the models that are mentioned in this chapter.

Around the same time that complexity science took off, Wolfram had an epiphany. He was studying a set of computational systems called cellular automata. Cellular automata are based on simple logical rules. One of the most famous cellular automata is "Conway's Game of Life," created by John Horton Conway. This game appeared for the first time in 1970 in *Scientific American* in a column named "Mathematical Games."

The game takes place on a square grid with black and white cells. Each cell interacts with the eight cells around it (its neighbors) at each turn. Conway came up with a set of four rules for interaction:

1) Any live cell with fewer than two live neighbors dies (referred to as underpopulation or exposure).

2) Any live cell with more than three live neighbors dies (referred to as overpopulation or overcrowding).

3) Any live cell with two or three living neighbors lives, unchanged, to the next generation.

4) Any dead cell with exactly three live neighbors will come to life.

Okay, this might sound somewhat abstract. What exactly does the Game of Life look like? You can find a web app version of the game in this link.[16] Let's start with some random placements of colored cells, shown in Figure 12-3.

---

[15] www.wolframalpha.com/

[16] bitstorm.org/gameoflife/

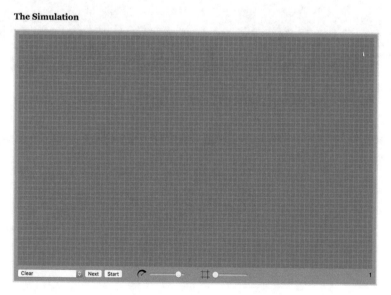

Figure 12-3. *Game of Life, round 1*

Right now we're in round 1 of the game. What happens in the next round? Well, we
have to follow the rules, so in the next round the cells simply die—they all suffer from
underpopulation. See Figure 12-4.

The Simulation

Figure 12-4. *Game of Life, round 2*

Okay, so that's not a very interesting example. But what happens if we change the initial pattern, so that some cells are clustered together? Let's try the exploder pattern for four rounds. See Figure 12-5.

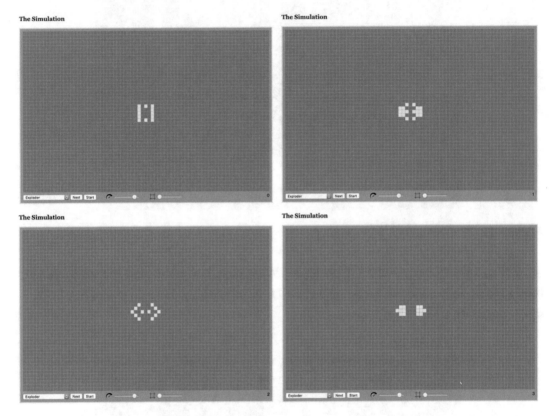

Figure 12-5. *More interesting patterns produced by Conway's Game of Life*

This pattern is clearly more interesting! What happens if we combine the patterns, as in Figure 12-6?

The Simulation

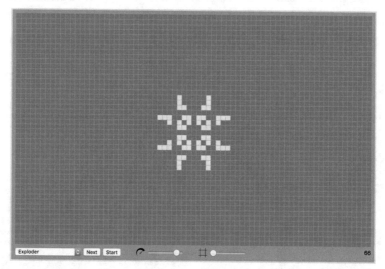

Figure 12-6. *Another pattern generated by the Game of Life*

There are many more patterns you can play around with. The main point of the Game of Life is that using four simple rules, you can get lots of different complicated and unpredictable behaviors. Conway's Game of Life has been proven to be Turing-complete. The term "Turing complete" relates to Alan Turing (1912–1954), the father of computer science, and it refers to an automaton (such as this one), that, in theory, has all the computational capabilities of a computer. This is an extremely interesting property, since it means that such a simple game can essentially compute all kinds of programs. In theory, it could be used to compute anything that we can use a modern computer for.

What Wolfram noticed was that the simplicity or complexity of the rules was not predictive of the simplicity or the complexity of the patterns that would emerge. Very simple rules in a cellular automaton could generate very complex patterns. For example, Wolfram was obsessed about Rule 30 in cellular automaton that could generate very complex patterns. Indeed, these patterns seemed to go on forever, without any way to predict what would happen next. Wolfram even used this cellular automaton as a random number generator in his software Mathematica.

Rule 30 in cellular automaton is depicted in Figure 12-7. This is an example of one-dimensional cellular automaton. You start at the top row, and the rules determine what the color of the cell below will be. So, you see that the middle cell in the top row is black

with two white neighbors. Hence, the cell below it is going to be black as well. The cells
next to it, in the first row, each are white, with one black neighbor (the black neighbor
being the black cell in the middle). Hence, the cell in the row directly below will be black.

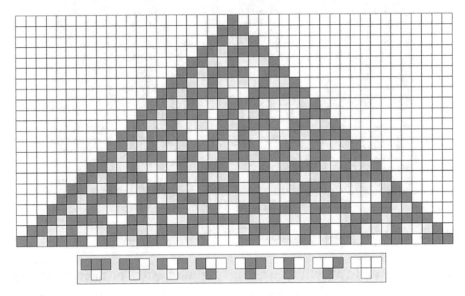

Figure 12-7. *Rule 30 in cellular automaton[17]*

Okay, this might not look impressive, but check out what happens after a larger
number of iterations, shown in Figure 12-8.

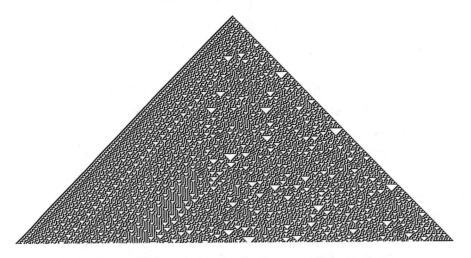

Figure 12-8. *Rule 30 in cellular automaton after multiple iterations*

---

17 _

There is no discernible pattern of evolution. There are some shapes that appear in different regions, but it is not really possible to predict the exact evolution of the automaton. Figure 12-9 shows how different this cellular automaton is from other rules, which produce simpler patterns.

Rule 110   Rule 150   Rule 54

Figure 12-9. *Other types of cellular automata*

It's worth noting that the people behind the creation of cellular automata were also the same people from the Los Alamos laboratory—von Neumann and Stanislaw Ulam.

Wolfram made the observation, similar to the ones made by complexity theorists, that from simple rules, complex phenomena might emerge. Many molecules together can create cells. The binary firings of cells of neurons give rise to consciousness. The simple interactions in the bartering of two merchants grows to become a global economy sophisticated enough to produce spaceships to travel to other planets. Wolfram proposed that the "new kind of science" would be based on computer simulations, where physical systems would be based on computational analogues, just like cellular automata. However, while Wolfram was very self-congratulatory of his discovery, and rather more derogatory about the more traditional approaches, he never really explained how it would take place. Again, cellular automata seemed to have many similarities with systems in the real world. But how could we use one to predict the other?

So, how does all this tie to uncertainty? I believe that one of the issues behind complexity theory was that, for many scientists, the vision behind it was fundamentally tied to the original vision set out by the reductionist approach that was established during the Enlightenment. Science had, for a long time, had great success explaining natural phenomena through analytical equations that were an accurate approximation of the natural world. The most famous example is Newton's law of gravity.

However, as you saw earlier, even Newton himself couldn't deal with the madness of men—for example, the emergent social phenomenon of an economy, a financial structure, and the subsequent frenzy that ensued during the bubble of the South Sea company. Those phenomena eluded scientists for a long time, and they still do, to a large extent. Complexity theorists believed that by using models, like cellular automata, with properties similar to real systems, they could come up with similar formulas or models that (much like Newton's law of gravity) could predict the paths of some of the phenomena discussed earlier.

For a scientist, this is more or less the ultimate goal. While complexity scientists rejected reductionism and believed emergence to be something unexplainable, they were essentially looking for the same kind of clarity that reductionists had achieved through models of the physical world. These are the kinds of models that provide the smallest degree of uncertainty. The laws of mechanics and chemistry give us great confidence as to the result of two objects colliding, or a chemical reaction. If we had models or methods that could make these kinds of predictions for all kinds of other complicated systems, then we would be credited with discovering the most important scientific breakthrough of all time.

I discussed earlier how Monte Carlo simulations were employed by physicists in Los Alamos, when they couldn't figure out exact solutions to some of the problems they were examining. They had hit an uncertainty limit, which was impenetrable by traditional methods. Hence, they had to apply other approaches that would explicitly incorporate uncertainty, in order to try and control it—know thy enemy, in other words. Complexity theorists essentially believed that they could get away with it. However, unwittingly, the methods they employed revealed to a large extent the challenges they faced and elucidated why their vision might not be accomplishable after all.

One of the most popular tools employed by complexity theorists was agent-based simulation. Let's say that you want to model the economy of a city using an agent-based model. Think of all the different kinds of uncertainty you would face. It would be impossible to completely simulate each individual's preferences and thought processes. Even if you were able to track each person 24/7 through a variety of data collection methods, you would still not be 100 percent certain about the decisions that every single individual might make. You would have to make many simplifications in the model, which would only add to the uncertainty.

Complexity theory borrowed and extended chaos theory's idea that small effects have disproportionate results. Indeed, the whole theory behind cellular automata is that simple rules, and simple interactions, can give rise to widely different phenomena.

Coming back to the example of the agent-based model we're building, we have already accepted the fact that there will be a high degree of uncertainty in many of the parameters of our model. By compounding all kinds of uncertainty, on top of this kind of sensitivity to the different conditions within a system, we get a huge number of potentially different paths that the system can take at any given moment. You will have probably heard various kinds of arguments as to what would have happened in history if some other event had or hadn't taken place. For example, how would history have unfolded if the German army had made some different strategic decisions and ended up winning World War II? How would have humanity have progressed if Einstein, for some reason, hadn't managed to come up with the theory of relativity?

You've probably thought about such examples in your own life. How would your life have unfolded if you had made different decisions? Furthermore, how would the lives of all the other people whom you have met—or haven't meet—have unfolded if you had made different decisions? This question has been the topic of many popular movies over the years, such as *Sliding Doors* and *The Butterfly Effect*.

When we try to simulate the economy through an agent-based model, we are trying to do exactly the same thing, but the task is exponentially more difficult. Uncertainty is compounded through all those interactions, and we can soon forget about creating anything that has predictive value.

Learning from Complexity: The Limits of Computation Are the Limits of Uncertainty

"Man is not born to solve the problem of the universe, but to find out what he has to do; and to restrain himself within the limits of his comprehension."

—Johann Wolfgang von Goethe

Throughout history, humanity has always shifted the paradigms around the limits of uncertainty. Before the renaissance, uncertainty was ruled by supernatural forces, gods and luck. Trying to study it was useless, as man was defenseless against the hand of fate. The work of people like Pascal, Cardano, Jacob Bernoulli, Gauss, and de Moivre brought probability to the fore. Graunt, Halley, and Quetelet brought measurement and life expectancies into the front and showed that humankind is not wholly defenseless against uncertainty.

During the Victorian era, the approach championed by people like Galton, Quetelet, and William Jevons was that of constant and careful measurement. You encountered Galton and Quetelet in a previous chapter. William Jevons (1835–1882) was an

English economist who, in his book *The Theory of Political Economy,* championed a
mathematical approach to economics and championed the approach of utility as a
notion of the measure of value. The notion that the value of a good or service can be
described by numbers was revolutionary, and it is a clear example of the paradigm that
existed during that era.

Before the Victorians, people who played a key role in the development of
mathematical tools for uncertainty, like Bernoulli and Laplace, believed that all events
could be attributed to causes. The only reason we don't have a complete picture
is because we don't have complete information. Essentially, they believed that all
uncertainty was epistemic. Laplace wrote in his *Essai Philosophique sur les Probabilité*s:

> "All events, even those which on account of their insignificance,
> do not seem to follow the great laws of nature, are a result of it just
> as necessarily as the revolutions of the sun."

This paradigm was abandoned after World War I, when the horrors of war
demonstrated that it is impossible to know everything and destroyed the hope
that uncertainty will eventually be obliterated. WWI was harrowing proof that,
really, we didn't know humankind at all, nor what it's capable of. Sigmund Freud's
theories exposed the irrationality of humans, and their violent, sexual, and repressed
unconscious which—Freud believed—is the driver of actions. Albert Einstein
demonstrated that Euclidean geometry is not an accurate description of our world.
Heisenberg discovered the uncertainty principle, demonstrating that complete
knowledge at the most granular level is impossible. John Maynard Keynes, in his 1936
masterwork, *The General Theory of Employment, Interest and Money*, voiced his opinion
against the strict application of probability for economic matters:

> "We are faced at every turn with the problems of Organic Unity,
> of Discreetness, of Discontinuity—the whole is not equal to the
> sum of the parts, comparisons of quantity fail us, small changes
> produce large effects, and the assumptions of a uniform and
> homogeneous continuum are not satisfied."

Keynes, in his theories, saw prediction of the future state of the economy, as well as
uncertainty, as a key economic variable. Households and businesses make decisions
based on predictions of what the future will look like. Decisions have an impact in
the world, changing the economy, and then creating new conditions, where old

opportunities might no longer be relevant. This led him to promote an interventionist policy, in contrast to the *laissez-faire* approach championed by other economists, who believed that an economy left to its own devices will end up in a stable equilibrium.

Similar opinions were voiced by Frank Knight (1885–1972), an economist of the famous Chicago school of thought, who objected classical economic theories that assumed that decision making takes place based on perfect information. His Cornell University PhD thesis, which was published in 1921, was the first work to deal explicitly with decision making under uncertainty.

The advent of the computer era, alongside the scientific discoveries during World War II, gave rise to a renewed enthusiasm about measurement and uncertainty, but also a more conscious understanding of the complex interactions that restrict our ability to predict with complete certainty the direction of a system. Neumann's Game theory, chaos theory, and complexity theory all exemplify this trade-off. However, complexity theory was combined with a Victorian-style enthusiasm for being able to explain everything. Unfortunately, they failed to explain or, even to predict the state of a system.

We can say that the current enthusiasm about data and measurement is the Victorian-era on steroids. Our current era tests once again the limits of prediction and measurement. With smart devices measuring everything about our lives, from sleep patterns to steps taken, it's reasonable to expect that the future lies in using predictions for everything, from our health to finding romantic partners (in fact, it already does—OkCupid was co-founded by Christian Rudder, a Harvard Mathematician who later wrote a book called *Dataclysm*). The success of machine learning, in spite of it lacking theoretical rigor, also demonstrates the utilitarian spirit of our times. Complexity theory tried to explain everything in a beautiful way, but didn't produce very useful tools, and it has been cast aside for a world of tools and methods that are simply more useful.

So, is complexity theory just a passing academic curiosity? Arguably not—complexity science gave us various useful insights.

First, the qualitative study of systems is valuable, even if it is not as informative as quantitative predictions. Understanding that systems are similar can give rise to new ideas and valuable theories.

Secondly, many tools that were championed by complexity science, like agent-based modeling, can actually be very useful. A carefully crafted agent-based model can be highly informative, even predictive, if it's done right. However, there are no general solutions there, as to what is the best model and how it should be built. If complexity theorists hoped to come up with a set of magic formulas to explain the world, then they failed. But if someone puts in the effort to build a model and is honest as to the assumptions made, then the model can yield useful results.

Agent-based models have been used successfully in various settings However, each model is different, and some of them might look overtly too complicated to the extent that you end up wondering whether the actual model ends up being nearly as complicated as the phenomenon being studied. However, we have to be explicit about our limits, especially with regard to uncertainty. You can choose to ignore uncertainty but it is always going to be there. The best thing you can do is to recognize it.

One of the reasons that data science approaches have dominated the area of uncertainty is that with enough data, we can get pretty accurate predictions and good models around a narrow set of questions. So, while building a simulation of a complete economy is difficult, using lots of different data sources to predict certain economic indicators can end up being much easier. Obviously, there is no rule saying that you can't combine approaches and many data scientists actually do that. Quite often, using a combination of methods, for example, agent-based modeling with machine learning, can yield more powerful results than using one method alone.

Finally, there is another fundamental reason as to why we might never be able to completely remove all uncertainty in complex systems. This relates to one of the most fundamental problems in computer science: the P=NP problem. Problems that fall under the P complexity class can be solved within a polynomial time. In practical terms, this means that we can find an optimal solution in a reasonable amount of time.

However, for other problems, the complexity class is NP, which stands for non-deterministic polynomial time. For these problems, no algorithm exists that guarantees a solution in polynomial time. In practical terms, this means that it can take a very long time to find a solution, and finding the optimal solution may be beyond practical means. However, the solution of an NP problem can be verified in polynomial time. So, while finding solutions might be difficult, verifying them is easy.

The P=NP conjecture asks whether every problem for which the solution can also be verified in polynomial time (that is, relatively quickly) also has a polynomial solution. In simple words, it asks whether every hard problem (to which a good solution might be impossible to find) can be reduced to a problem where a solution is easy to find.

So, why is this important? Many of the most important scientific questions and challenges are problems that relate to the NP class. Problems in genomics, like optimal threading procedures, are NP. Problems in optimization for important common tasks, like vehicle routing, or arranging classes in a university, are also NP. Cryptography has been based on the non-polynomial difficulty of solving cryptographic ciphers.

If we could prove that P=NP exists, then the world would look very different. It would be easy to calculate the optimal solutions to all sorts of problems, from transportation, to telecommunication networks, to genomics. However, there is much more we could do. If P=NP, then we could discover the optimal model for every occasion. We would even be able to create a model representation in the world.

In fact, this might be a great way to predict future states. So, why don't we do it? Let's assume that we had access to the laws of physics to such a degree of accuracy that we could build a simulation from the subatomic level up to the social level. We could build a completely accurate model of our universe, simulate it in a computer, and then predict with complete certainty what the next state of the world will look like.

Let's assume that we did indeed possess this knowledge. What would stop us from doing that? There are a couple of issues. First of all, the kind of computational power required to do this might exceed the computational power that can be provided by the universe, even if we were trying to utilize every atom as a binary processing unit. But even if this was not the case, because, for example, we have some incredible new technology like quantum computers, we would still need huge amounts of computational power, which is practically impossible to find.

The way to reduce the computational requirements is to find more efficient ways to represent parts of this model. So, instead of simulating every part of the system, we can use a more efficient mathematical model in order to achieve the same result. However, those mathematical models are likely of the NP class, which means that we still won't have the computational resources to do it.

Indeed, it looks as if nature itself has placed limits on our computational capabilities. That is why it is very likely we will always have to use abstractions, heuristics, and approximations in our models. If we want to predict the future or understand the state transitions of a system, we might never be able to reach the romantic ideal expressed by people like Laplace, where every bit of uncertainty is epistemic, and the only thing separating us from complete knowledge is just *more* information. Aleatoric uncertainty might be an inherent part of nature.

However, this is not necessarily a disaster. While the simulation of the universe might be impossible, for practical purposes, we can find good enough solutions to most of the problems we are facing. In addition, advances in computation, and maybe even the advent of quantum computers, will make this process even more efficient. Hence, while completely lifting the veil of uncertainty might be out of our reach, it might not be necessary for most practical applications.

The P=NP problem is one of the seven Millennium Prize Problems selected by the Clay Mathematics Institute, each of which carries a US$1,000,000 prize for the first correct solution. So far it goes unsolved, and we currently don't know whether a solution even exists. However, it is likely that P≠NP, and that this is one of the bounds of certainty that we have to learn to live in.

CHAPTER 13

Uncertainty in Us: How the Human Mind Handles Uncertainty

> *"We must content ourselves with the mystery, the absurdity, the contradictions, the hostility, but also the generosity that our environment offers us. It's not much, but it's always better than the deadly, defeatist certainty of the paranoid."*

—Philip K. Dick

Uncertainty is such a big part of life that our brains and minds have adapted to it over millions of years of evolution. For a large part of our lives, many of our thoughts and decisions are affected by, or are a response to, uncertainty.

Uncertainty can occur at different levels and timescales. We might face uncertainty about events that could happen in the next hour (e.g., will it rain?), or to uncertainty about events that will unfold over years (e.g., will my life be better if I move to a different city?). We face uncertainty about things we can control, like our decisions, but also about things that are completely outside of our control.

Uncertainty is a major cause of stress. This is a well-documented fact in medical literature.[1] The term "cognitive dread" describes a phenomenon where the actual anticipation of an event can be worse than the event itself. In a 2013 study,[2] researchers

---

[1] Grupe, D. W., & Nitschke, J. B. (2013). "Uncertainty and anticipation in anxiety: an integrated neurobiological and psychological perspective." *Nature Reviews Neuroscience*, 14(7), 488.

[2] Giles W. Story, Ivaylo Vlaev, Ben Seymour, Joel S. Winston, Ara Darzi, Raymond J. Dolan, (2013) "Dread and the Disvalue of Future Pain." *PLoS Comput Biol*, 9(11).

© Stylianos Kampakis 2023
S. Kampakis, *Predicting the Unknown*, https://doi.org/10.1007/978-1-4842-9505-2_13

faced the subjects with dilemmas. They could either choose real painful stimuli right now in the form of electric shocks, or hypothetical painful events in the future, like a dentist appointment. The vast majority of the subjects preferred to experience pain right now, rather than defer it. The mere expectation of pain in the future can actually be more discomforting than the pain itself. We prefer the certainty of experiencing discomfort now, rather than deferring it into the future. It's like the classic adage—sometimes you just have to rip the plaster (or the band aid) off.

In another experiment conducted in 2016,[3] researchers from University College London found that a small chance of getting shocked was significantly more stressful than a definite chance of getting shocked. The experiment involved 45 subjects, who played a computer game where the objective was to turn over rocks and guess whether there were snakes underneath them. If there was a snake under the rock, the subject would receive a mild electric shock.

The probabilities of finding a snake depended on the type of rock, and eventually the subjects learned the patterns, but then they changed as the game progressed. This caused a situation of uncertainty, which was controlled by the experimenters. They tracked stress by asking the participants to report it, but also measuring pupil dilation and perspiration (two common measures of stress response). A 100 percent chance of getting shocked was better than a 50 percent of *possibly* getting shocked. Stress directly correlated with uncertainty.

Simple, but smart, experiments like these clearly demonstrate that our brains are hardwired to respond to uncertainty. The reduction of uncertainty can also be a motivating force. In 1975, Charles Berger and Richard Calabrese proposed the uncertainty reduction theory,[4] which is applied in communication. According to it, uncertainty causes cognitive distress, so when two people meet, they seek information about the other party in order to reduce uncertainty about the other party's cognitive state and behavior.

Given that uncertainty is such a huge part of our lives, nature has imbued us with mechanisms to deal with uncertainty in different ways. The cognitive and brain sciences have recently discovered many of the fascinating ways through which our minds and brains deal with uncertainty, with rippling effects into our lives and society.

---

[3] www.ucl.ac.uk/news/2016/mar/uncertainty-can-cause-more-stress-inevitable-pain

[4] Berger, C. R., Calabrese, R. J. (1975). "Some Exploration in Initial Interaction and Beyond: Toward a Developmental Theory of Communication". *Human Communication Research*, 1, 99–112.

Uncertainty and Our Mind

"Our comforting conviction that the world makes sense rests on a secure foundation: our almost unlimited ability to ignore our ignorance."

—Daniel Kahneman, *Thinking, Fast and Slow*

Probably the most influential psychologists in the context of uncertainty are Daniel Kahneman (1934–) and Amos Tversky (1937–1966). Kahneman and Tversky studied the way people make decisions, and in doing so, they reshaped some of our long-held beliefs about the way that humans form conclusions and make choices about the future.

Kahneman and Tversky started working together in the 1960s, when both of them were professors at the Hebrew University in Jerusalem. Over the next decade, they conducted a series of experiments that culminated in what came to be "prospect theory." In 1979, they published a paper in the prestigious journal *Econometrica*, called "Prospect Theory: An Analysis of Decision Under Risk."[5] This paper has been cited more than 54,000 times. In this paper, they essentially destroyed the old paradigm of decision making in economics. Their work is considered to be the seminal paper in the field of behavioral economics.

Up to that point, theories of decision making were normative. They assumed that people made rational decisions according to expected utility. Kahneman's and Tversky's goal was to come up with a theory that describes how people *actually* make decisions— not how they *should* make them if they followed some notion of rationality.

The main observation that Kahneman and Tversky made was that people might treat losses and gains in different ways, depending on how a problem is being posed, even if the expectations are the same mathematically. Let's say that someone poses the following question to you. You have two choices:

> Choice A: A 50 percent chance of earning $200 or a 50 percent chance of earning nothing.

> Choice B: A 100 percent chance of earning $100.

From a mathematical viewpoint, they are equivalent. The expected value of choice A is $100. However, most people display risk-averse behavior and will prefer choice B to choice A.

---

[5] The full paper can be found here: scholar.princeton.edu/sites/default/files/kahneman/files/prospect_theory.pdf

Now let's say that someone presents the following choices to you:

> Choice A: A 50 percent chance of losing $200 or a 50 percent chance of losing $0.

> Choice B: A 100 percent chance of losing $100.

In this case, people are more likely to choose A. However, from a mathematical viewpoint, choices A and B have the same expected value.

That humans tend to be risk-averse was a well-known fact at that point in time. However, Kahneman and Tversky also discovered what is called the "failure of invariance." Changing the way that a question is posed also changes the outcome, while the mathematical expression is the same. People change to risk-seeking behavior when we are faced with uncertain losses.

This has tremendous implications for decision making in the real world. Kahneman and Tversky[6] posed the following problem. Imagine that a disease breaks out in some community and it is expected to kill 600 people. You are faced with a choice between two programs:

> Program A: 200 people will be saved.

> Program B: 1/3 probability that everyone will be saved and 2/3 probability that no one will be saved.

Most people (72 percent in the original study) choose program A, which is risk-aversive behavior. Kahneman and Tversky then asked the subjects to choose between the following two programs:

> Program C: 400 of the 600 people will die.

> Program D: 1/3 probability that no one dies, and 2/3 probability that everyone will die.

These two programs are exactly the same as before, they are just posed in terms of losses rather than gains. In this case, the majority of the subjects (78 percent) chose program D. By simply changing the way that the question is asked, they changed risk-averse subjects to risk-seekers.

---

[6] Kahneman, D., & Tversky, A. (1984). "Choices, values, and frames." *American Psychologist,* 39(4), 341-350.

This behavior was not consistent with the assumption that was being made up until that point that people follow strict rationality. Also, people are not risk-averse. Instead, they are loss-averse. Tversky writes: "It is not so much that people hate uncertainty—but rather, they hate losing."

At the heart of the problem is the reality that the world is complicated. Our mental and cognitive capacity is limited, and quite often we have to make decisions quickly. Our minds employ heuristics in order to save resources, which, as a side effect, creates biases in decision making. In an article published in the journal *Science* in 1974, called "Judgment Under Uncertainty: Heuristics and Biases," Tversky and Kahneman describe how humans use various heuristics to come up with decisions quickly[7] when faced with uncertainty.

One such heuristic is representativeness. When people are asked about the probability of something, they might ignore the base rate. Kahneman and Tversky conducted an experiment where they asked the following question to the subjects:

> "Tom W. is of high intelligence, although lacking in true creativity.
> He has a need for order and clarity, and for neat and tidy systems
> in which every detail finds its appropriate place. His writing is
> rather dull and mechanical, occasionally enlivened by somewhat
> corny puns and by flashes of imagination of the sci-fi type. He has
> a strong drive for competence. He seems to feel little sympathy
> for other people and does not enjoy interacting with others. Self-
> centered, he nonetheless has a deep moral sense."

Then they asked the subject what is the probability that Tom is on one of the nine major courses at university. Most subjects said that Tom is very likely to be an engineering student, even though engineering students were a small minority in their college. The subjects ignored that the *a priori* probability of someone being an engineering student is small, essentially failing to properly apply Bayes theorem. In short, we make snap judgments based on stereotypes.

---

[7] Tversky, A., & Kahneman, D. (1974). "Judgment under uncertainty: Heuristics and biases." *Science*, 185(4157), 1124-1131.

Another heuristic they found was the availability heuristic. Think quickly—are there more words in English with the letter K in the first or the third position? Most people will say there are more words with the letter K in the first position, at the start of the word, while in fact the latter is true. The reason this happens is that it is easier to recall words which start with the letter K, rather than recall words where the letter K is in the middle.

Finally, there's a heuristic—which is employed by sales people and marketing teams everywhere—that's called anchoring. A 50 percent discount of a product that drops the price from $200 to $100 looks more lucrative than a discount that drops it from $120 to $100. When a salesman negotiates a price, they want to start from a high price, so then any discount will end up looking lucrative to the buyer. Anchors can even work between things that have no association with each other. Kahneman and Tversky asked subjects to guess the percentage of nations that were African nations. However, before asking the subjects, the participants observed a single round of a roulette game, which was programmed to stop at either 10 and 65. The result of the roulette wheel was enough to change their estimates, with people who had seen 10 reporting lower percentages from those who had seen 65.

In 1961, Daniel Ellsberg proposed that humans are characterized by ambiguity aversion,[8] which was later confirmed by Kahneman and Tversky. Put simply, people just dislike uncertainty. However, as Kahneman and Tversky discovered, we employ various rules and strategies in order to deal with it, which makes for a more complicated behavior pattern than anyone had expected up to that point. Kahneman proposed that our minds work to two different systems:[9] System 1 (the fast mode) and System 2 (the slow mode). System 1 is efficient, but makes extensive use of heuristics and biases, hence many times it makes mistakes. System 2 is slow and conscious, so it's more accurate, but inefficient. System 1 is our automatic pilot and the default mode for most of our day (e.g., when driving from back home from work), but we engage in System 2 thinking when the need arises.

The heuristics and biases that have evolved are often portrayed in a negative light, but they have come as the result of millions of years of evolution, and they can serve us well—even save our lives—in many other situations when making decisions quickly is of paramount importance. It's just that at times they can be the source of bad decisions.

---

[8] Ellsberg, D. (1961). "Risk, ambiguity, and the Savage axioms". *The Quarterly Journal of Economics,* 643-669.

[9] Kahneman, D. (2012), *Thinking Fast and Slow.*

While the work conducted by Kahneman and Tversky concerned economics, it's clear that that uncertainty plays a key role in the way our minds work. This has huge implications for society, as well as the economy. Daniel Kahneman, in his book *Thinking, Fast and Slow*, described how entrepreneurs are "delusional optimistic." When faced with uncertainty, entrepreneurs, by nature, believe that the future will be better than it might really be. Sometimes, being a bit delusional can be a good thing.

One of the most important researchers in this area is Richard Thaler, who was also awarded the Nobel prize in economics in 2017 for his contribution to behavioral economics. In his 2008 book *Nudge,* he outlined his vision for how those heuristics and biases can be used in order to nudge citizens unconsciously toward desired behaviors. For example, most people will go for the default choice in anything. In countries where the default option is for someone to become an organ donor after death, there are more donations than in countries where someone has to explicitly state they want to become organ donors. Hence, by setting donation to be the default, a country can easily increase the availability of organs for transplants. This is exactly what the British government implemented in 2020.

An important point to remember is that even when uncertainty is not the central theme, it can still have a huge impact on our lives through its indirect effects on other aspects of our minds and biology.

Another set of theories for how the mind operates postulates that our cognitive systems function according to the principles of Bayesian inference. Some of the most prominent researchers in this area are Tom Griffiths, Josh Tenenbaum, Nick Chater, and Charles Kemp. However, related theories had been proposed as early as the 1860s by Hermann Helmoltz (1821–1894). So, how exactly can we use Bayes in cognition? The Bayes theorem provides a way to merge prior knowledge with new data. Let's say that you are playing tennis and you see the ball coming toward you. Based on your experience, you have some prior knowledge of where the ball usually lands. This is also influenced by the experience you have about your opponent, and the opponent's position in the court.

After your opponent hits the ball, your brain starts collecting data about the actual trajectory of the ball. So you combine any prior information you had about the most likely spots for the ball to land, along with updates on the current trajectory, and build a more accurate estimate with each passing moment. This can be easily described using Bayes formula:

$$P(sensory\ input) = \frac{P(state)P(state)}{P(sensory\ input)}$$

There's evidence that our cognitive systems integrate sensory information in just such a way.[10] This can easily be seen in optical illusions such as the one in Figure 13-1. If you're like most people, you will see four bumps and two depressions. Usually light sources (e.g., the sun) are above us, so we have a strong prior for that assumption. Each circle, as you can see, is painted with a shadowy area either at the top or the bottom. If the shadowy area is at the bottom, then the circle looks like a bump, because this would explain the shadow. If the shadowy area is at the top, then the circle is seen as a depression, because only this would explain the existence of a shadow from a light source that shines from above. If you reverse the image, you will see four depressions and two bumps, which is the opposite of the previous patterns.

[10] Kording, K. P. & Wolpert, D. M. "Bayesian Integration in Sensorimotor Learning." *Nature* 427, 244-7 (2004)

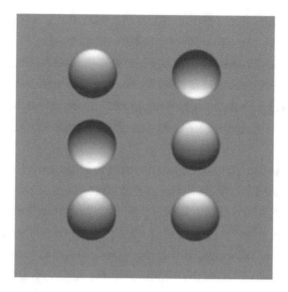

Figure 13-1. *The shape-and-shading illusion[11]*

Another good example of how our brain generates our version of the world is *pareidolia*. Pareidolia describes the phenomenon of seeing familiar patterns in vague visual input, e.g. observing human faces in the clouds.

There are many publications in this field, but it doesn't go without its critics. The main criticism against Bayesian explanations of cognition is that, simply because Bayes theorem and models can explain experimental results, it doesn't mean that our minds are necessarily operating in this manner. As Josh Horgan wrote in an article in *Scientific American*:

> "Moreover, the Bayesian-brain thesis can be boiled down to a dubious syllogism: Our brains excel at certain tasks. Bayesians programs excel at similar tasks. Therefore our brains employ Bayesian programs."

While the debate around the soundness of the Bayesian approach to cognitive modeling might never end, there is a new theory that has emerged recently that provides an explanation of how our minds and other biological systems make inferences and decisions under uncertainty. This is Friston's "free energy principle." Karl Friston is a neuroscientist at University College London, and is also (at the time of writing) the most cited neuroscientist in the world.

---

[11] mlg.eng.cam.ac.uk/zoubin/papers/WolGha06.pdf

Something fascinating that's explained by Bayesian theories of cognition is that our minds are not simple passive observers of uncertainty. Instead, our minds follow a generative model. In the illusion shown in Figure 13-1, your brain didn't simply register the information passively, and you didn't have to spend a significant amount of time to think about the illusion. Your mind came up with a very quick interpretation of the circles and the shadows. You are generating a version of the world, actively interpreting results.

Friston takes this a step further. He argues that the minimization of uncertainty is essentially the goal of all life. His free energy principle places uncertainty at the center of all intelligent behavior. Friston bases the principle on some basic postulates. All biological organisms tend to resist disorder, and in order to do that they have to match their sensory input with predictions about the world. There are two ways to do this—by either updating the theories about the predictions (what to expect from the world), or by taking action to minimize surprise. The term "free energy" arises from the Bayesian treatment of the problem. The free energy is a Bayesian quantification of uncertainty, between what an organism expects and what sensory input it receives. The brain, and hence the mind, takes an active role in generating theories about the world, but also affecting it.

This is a very high-level, broad explanation of the theory. Friston's body of work on this theory spans more than a decade and it can seem esoteric and impenetrable, as it combines concepts from several different fields—from neuroscience to Bayesian statistics. Friston has also promoted his theory as an explanation to all phenomena related to the brain and the mind, from the function of neurotransmitters to mood disorders.[12] There are plenty of people who have voiced their inability to understand his body of work. For example, in a research digest Peter Freed wrote:[13]

> "At Columbia's psychiatry department, I recently led a journal
> club for 15 PET and fMRI[14] researchers, PhDs and MDs all, with
> well over $10 million in NIH[15] grants between us, and we tried to

---

[12] Clark, J. E., Watson, S., & Friston, K. J. (2018). "What Is Mood? A Computational Perspective." *Psychological Medicine,* 48(14), 2277-2284.

[13] Peter Freed (2010) Research Digest, "Neuropsychoanalysis: An Interdisciplinary Journal for Psychoanalysis and the Neurosciences," 12:1, 103-106

[14] PET stands for positron emission tomography, and fMRI for functional magnetic resonance imaging. Both are technologies to map brain activity.

[15] www.nih.gov/grants-funding

understand Friston's 2010 *Nature Reviews Neuroscience* paper—
for an hour and a half. There was a lot of mathematical knowledge
in the room: three statisticians, two physicists, a physical chemist,
a nuclear physicist, and a large group of neuroimagers—but
apparently we didn't have what it took. I met with a Princeton
physicist, a Stanford neurophysiologist, a Cold Springs Harbor
neurobiologist to discuss the paper. Again blanks, one and all."

There is even a Twitter account that parodies the difficulty of understanding the
theory.[16] Regardless of whether Friston's theory proves to be the dominant paradigm in
brain and mind sciences or not, it demonstrates one point very clearly: Uncertainty is
at the core of our existence, and even when we don't realize it, our actions and bodily
functions, down to the microscopic level of neurotransmitters, are affected by the impact
of uncertainty over the ages. Friston's theory places uncertainty as one of the primordial
forces of the universe, and in the epicenter of our existence.

Uncertainty and Our Brain

*"I believe that things just happen in life, and pretty much after the fact, we
make up a story to make it all seem rational. We all like simple stories that
suggest a causal chain to life's events. Yet randomness is ever present."*

—Michael S. Gazzaniga

Decision-making has shaped the way our minds work, but how is this reflected in the
brain? Our brain's basic communication mechanisms are the neurotransmitters. The
neurotransmitters are chemical substances that are exchanged from one neuron to
another through chemical pulses. Neurotransmitters play a role in everything we do,
from moving our bodies, to our feelings.

One of the most important neurotransmitters for cognition is dopamine. Dopamine
plays a huge role in reward. The *striatum* is a structure inside the brain that is often
called the brain's "reward center." It receives messages from many other places in the
brain, relating to motivation, decision making, and planning. For people addicted to
substances or habits like gambling, the striatum floods with dopamine every time they
give in to their addiction.

---

[16] twitter.com/farlkriston?lang=en

It was recently discovered that the striatum performs another important function. It not only rewards certain actions, but it also calculates probabilities of consequences.[17] The striatum is activated at its highest when the odds approach 50 percent. That is, at the point of maximum uncertainty between good and bad consequences. This has been found to be a key mechanism in gambling addiction. Gambling addicts are sensitive toward the dopamine released as a result of uncertainty.[18] Our brains are hardwired to respond to uncertainty, and gambling is a brain hack that can prove disastrous for those people whose brains are sensitive.

However, dopamine is only part of the picture, as there are other neurotransmitters involved in dealing with uncertainty. In a research paper published in 2016 in the *PLOS* journal,[19] researchers from University College London identified that besides dopamine, the neurotransmitters noradrenaline and acetylcholine play a crucial role in decision making under uncertainty. Noradrenaline regulates our estimates of how unstable the environment is and acetylcholine balances the attribution of events to chance. The role of dopamine is to push us to act on our beliefs about uncertainty. Other pieces of research have also indicated that the neurotransmitter norepinephrine plays an important role in uncertainty, by being used as a signal of unexpected uncertainty.[20]

Researchers from Yale University discovered that our brains are more activated and become better learners in situations where uncertainty is involved.[21] If our brains learned *all* the time, then we would face two adverse outcomes. First, our brains would have absorbed all kinds of wrong information and spurious correlations. Just because a train got delayed once, it doesn't mean you have to learn that it will be always delayed.

---

[17] www.theguardian.com/commentisfree/2016/apr/04/uncertainty-stressful-research-neuroscience

[18] Linnet, J., Mouridsen, K., Peterson, E., Møller, A., Doudet, D. J., & Gjedde, A. (2012). "Striatal Dopamine Release Codes Uncertainty in Pathological Gambling." *Psychiatry Research: Neuroimaging,* 204(1), 55-60.

[19] Marshall, L., Mathys, C., Ruge, D., de Berker, A. O., Dayan, P., Stephan, K. E., & Bestmann, S. (2016). "Pharmacological Fingerprints of Contextual Uncertainty." *PLoS Biology,* 14(11), e1002575.

[20] Angela, J. Y., & Dayan, P. (2005). "Uncertainty, Neuromodulation, and Attention," *Neuron,* 46(4), 681-692.

[21] news.yale.edu/2018/07/19/arent-sure-brain-primed-learning

Secondly, we would spend more resources than we need to. As the research in heuristics and biases demonstrated, the human brain will spend more resources only when needed, and for most problems it will resort to the "fast mode" of thinking. Hence, knowing when to learn and when not to learn is crucial.

What the research from Yale demonstrated was that when faced with uncertainty, learning improves. This also makes sense from an evolutionary perspective, given that this is when the odds are higher. A situation of high uncertainty usually presents to us a dilemma: It is either a situation with large potential gains or large potential losses. Learning how to separate the two could have meant life or death for early humans.

Even our two brain hemispheres—left and right—have been found to be specialized according to the ways they approach uncertainty and vagueness. It is well known, since the pioneering studies of Michael Gazzaniga (1939–) and Roger Sperry (1913–1994), that each brain's hemisphere performs different functions. Gazzaniga and Sperry studied patients with split-brain syndrome in the 1960s and 1970s. In split-brain syndrome, the patients' *corpus callosum,* which connects the two hemispheres, had been cut off. In various experiments, the patients demonstrated behavior as if there were two selves in one.

For example, when the young patient Paul S was asked what the right hemisphere wanted to do when he grew up, he replied "an automobile racer," but the left hemisphere responded "draftsman." Another patient was pulling his trousers down with one hand (controlled by one hemisphere), while the other hand was pulling his trousers up (controlled by the other hemisphere). Similarly, another patient was trying to choke his wife with one hand, while the other hand was trying to stop him from doing that. In popular culture, we believe that the right hemisphere is more expressive, artistic, and holistic, and the left more rational and the center of language. While this is a neat simplification, it's not entirely inaccurate.

The left hemisphere is focused on resolving uncertainty, while the right hemisphere is focused on resolving inconsistency. The left hemisphere hates uncertainty and will come up with any sort of explanation in order to justify what happened. In a famous experiment,[22] a patient with split-brain syndrome was shown two different images. The left hemisphere was exposed to an image of a chicken claw, and the right hemisphere

---

[22] Gazzaniga, M. S. (2000). "Cerebral Specialization and Interhemispheric Communication: Does the Corpus Callosum Enable the Human Condition?," *Brain* 123, 1293–1326. doi: 10.1093/brain/123.7.1293.

was exposed to an image of snow. He was then asked to select images with each hand that would go with the image being shown. With his left hand (controlled by his right hemisphere) he selected a shovel, which matched the snow scene, and with his right hand (controlled by his left hemisphere) he selected a chicken, which matched the chicken claw.

When asked by the experimenter why he chose the shovel with the right hand, the left hemisphere, which controls language, didn't have access to the information that the right hemisphere had seen the snow. The left hemisphere came up with the explanation that the shovel was chosen to clean out the chicken shed. In similar experiments, it was demonstrated how the left hemisphere will just come up with any kind of explanation in order to resolve uncertainty.

The right hemisphere seems to be responsible for detecting inconsistencies between theories. In a set of experiments, the left hemisphere can make quick inferences and become locked in them, even when it is incorrect.[23] It is the right hemisphere's job to detect inconsistencies and revise hypotheses about the world. Dealing with uncertainty in the best possible way requires a balanced approach between the two hemispheres.

Uncertainty also shows up in other places in our brain, like the debilitating condition of paranoid schizophrenia. Paranoid people suffer from a case of wrong beliefs. Think about how many of the things you do every day are based on implicit beliefs about the environment. For example, you believe that your spouse is not conspiring against you. You assume that the government provides public services which, at least in principle, have the citizens' best interests in mind (unless you live in an authoritarian regime). You believe that your business partner is not trying to backstab you. You believe that there is a sewage system running beneath the city you are in, even if you have never seen it, and that there are other continents, besides the one you live in, even if you've never been there yourself. You also believe that dragons are fictional creatures and exist only in fantasy books.

For people with paranoia, some or all of these beliefs will be different to yours. Other people might be out to get them, the government is controlled by malevolent non-human entities, and other continents do not exist; it's just a huge conspiracy trying to control our lives. Finally, dragons do exist, it's just that an evil wizard has transferred all of them to a different dimension.

---

[23] Marinsek, N., Turner, B. O., Gazzaniga, M., & Miller, M. B. (2014). "Divergent Hemispheric Reasoning Strategies: Reducing Uncertainty versus Resolving Inconsistency." *Frontiers in Human Neuroscience,* 8, 839.

What does all this have to do with uncertainty? Simply put, you can never be 100 percent certain about most things. You believe that there are other continents out there, simply because someone told you so, and this someone is a figure of authority (e.g., your parents). This is further reinforced by other authoritative sources (e.g., teachers). You have used multiple cues, quite often unconsciously, in order to reach the conclusion that this information is correct. However, in this process—strictly speaking—there are multiple objections that can be raised, relating to its soundness.

You saw in previous chapters how confirming and disconfirming hypotheses is something that has baffled philosophers and scientists since the dawn of time. Most people will intuitively use Occam's razor in situations such as the previous ones. The information you've been given is correct, because assuming otherwise would require you to make many over-complicated assumptions. Not so for a paranoid person. The theory that there is a complex conspiracy involving thousands of people in order to hide the truth that Atlantis exists is more plausible to them than simply assuming that none of these people would have any motive or any other reason to do this.

For most of us, most of the time, our brain automatically resolves uncertain scenarios without having to think too much about it. But some people fixate on the wrong things.

It's been discovered that, once again, it is the neurotransmitter dopamine that plays a huge role in this process. Dopamine decides where we focus our attention, and which details in a story are more salient. Dopamine dysfunction can cause delusions, including paranoia. The articles in a newspaper can seem to hide important messages, and the people around us are no longer the same people, but robotic impostors. In a paper about the neurophysiology of delusions, Corlett, et al. from Yale University[24] wrote:

> "[...] delusions (and their neurotransmitter basis) represent
> a failure to properly encode the precision of predictions and
> prediction errors; in other words, a failure to optimize uncertainty
> about sensory information. [...] the pathophysiology of delusions
> involves a misrepresentation of salience, uncertainty, novelty or
> precision (mathematically precision is the inverse of uncertainty)."

---

[24] Corlett, P. R., Taylor, J. R., Wang, X. J., Fletcher, P. C., & Krystal, J. H. (2010). "Toward a Neurobiology of Delusions," *Progress in Neurobiology*, 92(3), 345-369.

Everything, from our brains to our societies, has been shaped by the fight against uncertainty. When those systems, developed over millions of years of biological evolution, work in harmony, we thrive. When they do not, the consequences can be severe.

The same can be said about our social systems. Society has been created to a large extent as a response against all the uncertainties that nature imposes on us, from where we will get our next meal, to protecting ourselves against hostility and natural disasters. As we move forward into the 21st century, our tools for controlling uncertainty—either through prediction or through other kinds of engineering (e.g., financial instruments that control risk)—become ever more sophisticated. The technology you learn about in the next chapter presents the next stage of evolution for many uncertainty-based products, such as financial instruments.

CHAPTER 14

Blockchain: Uncertainty in Transactions

"Whereas most technologies tend to automate workers on the periphery doing menial tasks, blockchains automate away the center. Instead of putting the taxi driver out of a job, blockchain puts Uber out of a job and lets the taxi drivers work with the customer directly."

—Vitalik Buterin, founder of Ethereum

Think how many times you buy a product in your everyday life. This could be anything from a small transaction (e.g., a sandwich), down to a significant one (e.g., a new car). As a consumer, you are exposed only to one layer of complexity in this transaction: that of the buyer and seller. But think how many additional steps are required for the car to arrive at the dealership. A car is made of many components and consists of many materials, from metal to plastic to glass. The manufacturer has to transact with different companies to get access to those components. Then, the car has to be tested, and once the tests are passed, it has to be assembled in an assembly line. Each car that is being produced could be faulty, which might require additional tests. Then, the cars have to be transferred to different destinations all over the world, which can include the use of various transportation companies. The transportation companies have to use trucks and ships, either by buying them or arranging agreements with other companies that can carry their goods for them.

Buying a car is a miracle of the modern economy. You know that every time you visit the dealership, cars will be there. But the cars being there is conditional on a huge and long chain of trust, in which all the parties involved have to trust the other parties in everything: from the delivery of the goods to the quality of the goods.

© Stylianos Kampakis 2023
S. Kampakis, *Predicting the Unknown*, https://doi.org/10.1007/978-1-4842-9505-2_14

Now think about the money that you spend on buying the car. Money is embedded in our culture. You know you have to work to get money, and few people go about their lives never having worried about money. When you get a bonus at your job, you can't help but feel happy. Money—at least a version of it, and the act of transaction—has been around for as long as written history has been around, maybe longer. Hence, when you give money to the dealership to buy a car, you're not thinking too much about the complexity of the conventions involved behind money.

The monetary system has gone through various transformations. Ancient civilizations, like the Greeks and the Romans, used money made out of valuable metals. Other civilizations used other kinds of tokens as a medium of exchange. The gold standard that was used by some countries even as early as the 19th century[1] meant that a country had to back up the value of its currency in gold. The Bretton Woods agreement in 1944 established that other currencies were pegged to the U.S. dollar and the U.S. dollar was pegged to the price of gold. This was abolished in 1973, and the exchange rates were allowed to roam freely. The introduction of telecommunications into our lives also saw the introduction of plastic money through cards, in the second half of the 20th century. Now we can spend money buying something on the Internet, or travel on a metro system, without even having to take our card out of our wallet.

Inexorably, any monetary transaction is, at its heart, a transaction based on trust. When you perform such a transaction, the following assumptions are being made:

1) The money received is not counterfeit.

2) Both parties agree the currency being used is indeed a medium of exchange.

In order to conduct a monetary transaction, you need to ensure that the money is not "fake" and that both parties agree that the money is a valuable medium of exchange. When you receive your salary, you are happy, because you know that you use it to buy goods and services. If the currency you hold cannot be used for this purpose, you can exchange it with some other currency at a currency broker. Banks sit as intermediaries in this system, allowing the storing, creation, and exchange of money, and playing the role of a trusted authority. You don't have to worry that the bank will steal your money every time you store money or make a payment. The process is so intuitive you don't even have to think about it.

---

[1] England was the first country to adopt the gold standard in 1821.

However, someone did, and this changed the way we view money forever. In 2009, an entity called Satoshi Nakamoto published a paper called "Bitcoin: A Peer-to-Peer Electronic Cash System."[2] To this day, we still don't know who this person is, or whether it is a group of people instead of a single individual. In this paper, Nakamoto outlined how a currency could work without the need for a central authority. Money, Nakamoto argued, didn't require a central authority (e.g., a bank or a government), and it didn't require two parties to blindly trust each other. Money is a convention, and it can work as long as there is an algorithm in place that enforces trust. In the rest of the paper, Nakamoto explained how an algorithm combining elements of probability theory, game theory, and cryptography, based on a technology called "blockchain," could create this new kind of decentralized currency. In 2010, bitcoin was born.

It was the beginning of a revolution. The next few years saw an explosion of different types of cryptocurrencies[3] and different types of blockchains. The term "blockchain" on Google currently returns more than 300 million results. In 2019, the prices of cryptocurrencies sky-rocketed and blockchain became the hottest word in tech. But then the prices crashed, ending up in disaster for many investors. However, blockchain is still going strong, with many new applications coming out all the time. At the time of writing, all the big tech companies, from Microsoft to IBM to Amazon, are investing in blockchain. There are more than 300 cryptocurrencies being tracked by Coinmarketcap.[4]

In any transaction, monetary or otherwise, there is trust involved. Trust creates a condition of uncertainty. It is not possible for everyone in an economy to trust everyone else all of the time. Legal and accounting systems have been created since ancient times to assist in building trust, but they often fail. Blockchain is the next evolution of those systems. The next section explains why blockchain has been called the "Internet of trust."

The Internet of Trust

The problem that bitcoin was trying to solve was the creation of a decentralized, "trustless" currency. Bitcoin is decentralized because it does not require a central authority to function. It is trustless, because the exchange of the currency does not

---

[2] For the original paper, go to: `bitcoin.org/bitcoin.pdf`

[3] This is a term that refers to money based on blockchain technology. Common currency, like the USD, is referred to a "fiat currency."

[4] `coinmarketcap.com/all/views/all/`

depend on the beliefs or the reputation of each party. The system works because there is an algorithm behind it that makes it work, and this algorithm does not depend on a centralized authority.

Vitalik Buterin is a Russian-Canadian programmer. In 2013,[5] he wrote a whitepaper on GitHub explaining how the blockchain technology, which forms the backbone of bitcoin, could be used to create smart contracts. Ethereum was launched in 2015, and it is the single most important project, besides bitcoin, in blockchain development.

The insight that Buterin had was that, while the problem that Bitcoin was trying to solve was the decentralized and trustless exchange of currency, the *actual* problem that blockchain was trying to solve was trust. Smart contracts are an agreement between two parties that are verified and checked on the decentralized Ethereum network. A smart contract can include any kind of agreement and condition.

For example, a smart contract between a supplier and a buyer can be created so that as soon as the goods arrive, the buyer pays the supplier. A smart contract between an insurer and an insured party can be used so that as soon as a car is damaged, the insured receives compensation.

This can obviously be achieved with traditional technological means. However, the difference with Ethereum is that the verification of the smart contracts takes place in a decentralized, trustless, transparent manner. Blockchain promises the following advantages:

1) Decentralization: No single authority has the power to affect the system.

2) Transparency: The transactions are available for everyone to see and audit.

3) Immutability: Once a transaction is recorded on blockchain, it can't be changed.

Blockchain promises a decentralized and trustless world, where transactions can take place without intermediaries. But how does it all work?

---

[5] github.com/ethereum/wiki/wiki/White-Paper

How Blockchain Works

In technology, things often move toward higher layers of abstraction. Bitcoin, as an innovation, was focused on money and utilized blockchain technology. Blockchain, we found out, is a more general technology that can be used for any kinds of agreements, not only those including money. The broader category, which also encapsulates blockchain, is the distributed ledger.

Transactions are traditionally all placed in a ledger. Ledgers have been around since the beginning of bureaucracy. A distributed ledger is a ledger that is shared among all parties in a system. Hence, any transaction that takes place is being recorded by all parties, and there is a common history—from the first transaction to the last transaction—on which all the parties agree.

Why is this important? The original problem that Bitcoin tried to solve was how to avoid double spending in a digital currency. Let's say that there is a group of people who have agreed to use a digital currency, which is represented by computer files. What prevents a single party from copying a file a huge number of times, and then becoming a millionaire? Well, nothing apparently. Someone could easily copy the file, use a currency, but still be in possession of this currency. This is called the double spending problem. If Amara gives one token to Bob, but then she can still use this token (e.g., because she copied the file) to spend it in some other good or service, then Amara is double spending. Fiat currency avoids double spending in two ways:

1) Physical mediums of exchange (notes and coins) cannot be copied and forging is made deliberately difficult by the authorities.

2) Digital representations of money are protected by trusted authorities, which are the banks.

Bitcoin solved the problem in the following way. A ledger records all transactions from the first group of transactions (called the genesis block) down to the last one. Transactions are organized in blocks that include information about the sender, the receiver, and the hash of the previous block. When Amara sends one Bitcoin to Bob, a set of nodes in the system, called *miners*, check the ledger from the genesis block in order to ensure that Amara indeed possesses the required amount of currency.

However, things are not easy for the miners. In order to check the ledger, they have to solve a difficult cryptographic puzzle. The solution of this cryptographic puzzle requires the excessive use of electricity. This has a significant monetary cost. The first miner to find the solution is rewarded with Bitcoins. Once a solution is found, it is broadcasted to the rest of the network, adopted by the rest of the nodes, and then the whole cycle starts again.

I mentioned that a block contains a hash. A cryptographic hash function is a function that maps any number or letter to a fixed number of characters. For example, if you input the number 0 into SHA-256 (the hash function used by Bitcoin), the result would be the following:

07b01b0f4672f2bc58ef11132df4bc74a4e0dc9f2e07
ee5d9a0428d3836bc6cb

In Bitcoin, each block contains information about all the transactions that took place since the last block, but also the hash of the previous block. Solving the cryptographic puzzle means that you have to take into account the hash of the previous block, which depends on the block before that, which depends on the block before that, which all depend on the first block of transactions.

This dependency on the previous blocks means that if someone wants to game the system, they have to spend a huge amount of electricity to change the past. In 2018, it was reported that the global cost of mining Bitcoin in terms of electricity was more than the electricity consumed by the Czech Republic.[6] Attacking Bitcoin is not a valid economic proposition; it makes more sense to simply play along with the system. If someone wants to double spend in the Bitcoin blockchain, they need to control at least 51 percent of the computational power of the network. The further into the past a transaction is, the more difficult it is to change. Hence, consensus is enforced through incentives.

This protocol just described is called *proof-of-work*. Is the system perfect? Well, no. First of all, cryptocurrencies have been attacked in a 51 percent attack. In 2018, around $20 million were stolen through 51 percent attacks on smaller cryptocurrencies.[7] Smaller cryptocurrencies have smaller networks of nodes and are therefore more exposed to this danger. Secondly, the amount of electricity spent in proof-of-work has a huge environmental impact. The way that blockchain comes to an agreement is called a

---

[6] cointelegraph.com/news/bitcoin-minings-electricity-bill-is-it-worth-it
[7] thenextweb.com/hardfork/2018/10/23/cryptocurrency-51-percent-attacks/

consensus protocol. There are other consensus protocols besides proof-of-work that are promising to solve some of the problems it faces. For example, *proof-of-stake* has been promoted as an environmentally friendlier alternative, and there are other protocols such as *proof-of-elapsed-time* and *delegated-proof-of-stake*.

Are there decentralized ledgers that are not blockchains? Yes, there are. Some examples include Corda, IOTA, and HashGraph. Non-blockchain distributed ledgers do not require the storage of all transactions from the past until now.

Also, there are blockchains that are not public. While the public was fascinated by cryptocurrencies in 2018, private blockchains, promoted as solutions for the enterprise, started rising in popularity. Hyperledger Fabric is one such blockchain project. This private (or permissioned, as they are called) blockchain is open only to preapproved members. Enterprises can use those blockchains to get the benefits of blockchain technology, such as transparency and immutability, without having to use a cryptocurrency.

From Crypto-Anarchism to Drug Trafficking: The Unconventional Beginnings of an Interesting Technology

The rise of blockchain has an interesting story. Bitcoin tried to revolutionize the way we perform monetary transactions. However, for a long time, not many places were using it. One of the most popular places that it was used was the Silk Road (not the ancient Eurasian trade routes). The Silk Road was a drug marketplace on the dark web. The dark web is a network not indexable by search engines; it requires special software to access it.

So, how does an illegal marketplace operate on the Internet? How can someone pay a drug dealer without being found by the police? The answer is Bitcoin. Bitcoin, among other things, is pseudo-anonymous. Transacting parties do not need to use their real names. It is *pseudo*-anonymous because, as it was discovered later, it is possible to trace back transactions to individuals under some circumstances. However, early on, many believed it to be fully anonymous.

Bitcoin was Silk Road's cryptocurrency for a long time. Silk Road was busted by the FBI and Europol in 2014. This ironically helped popularize Bitcoin even further. Bitcoin also became a medium of exchange for those who wanted to hide money or avoid paying taxes. Hence, Bitcoin's popularity increased even further, but for the wrong reasons. However, it's also worth noting that many of the people involved in the early days of Bitcoin adoption were crypto-idealists, who really believed in Nakamoto's vision. These days, the popularity of Bitcoin has increased with many individuals as well as companies treating it as a store of value. The best example of this is Michael Saylor from MicroStrategy, who has invested company funds into Bitcoin as an investment.

Ethereum's arrival proved that blockchain has other uses besides the exchange of money. While smart contracts were supposed to be the way that transactions will take place in the future, the actual revolution came through Initial Coin Offerings (ICOs). Smart contracts allowed anyone to create their own cryptocurrency on the Ethereum blockchain. Cryptocurrencies created in this way were called tokens. Hence, token economies and ICOs were born.

The concept behind ICOs is as follows. A business uses a token as the currency of its platform. In order to purchase this token, you need to spend Ethereum or Bitcoin. An ICO is simply like an initial public offering (IPO), but it uses tokens instead of shares. Tokens are issued to the public, and then these tokens are purchased at either a fixed price or through some auction mechanism. The public purchases tokens because they either believe they will appreciate in value or because they want to use the tokens in the platform.

For example, Filecoin is a decentralized version hard drive. You can use Filecoins to store files in the Filecoin network, which is composed of other users. You can also earn Filecoins by allowing other users to use your free space on your hard drive. Tokens allowed the creation of what is called "token economies." Token economies allow the use of arbitrary incentives (e.g., using your hard drive space) to get rewards. Some examples, like the aforementioned Filecoin, could only exist on a blockchain.

But token economies aside, why would a company want to raise money this way? Raising funds this way is much easier than going through other means (e.g., traditional crowdfunding). Furthermore, the fact that blockchain is global means that anyone in the world can very easily buy those tokens in an instant. Cryptocurrencies make international transactions look easy. Finally, the increasing popularity of blockchain projects means that people were willing to throw tons of money at anything blockchain related.

In 2018, ICOs raised more than $7 billion.[8] Most of these projects went bust, which comes as no surprise, given that there were companies asking for millions of dollars without even having built a product. For a measure of comparison, in many cases, seed funding in most countries does not exceed $1 million, and even then it is likely that the company asking for money has built a product prototype. In 2019, ICOs became a toxic word, with most investors abandoning them.

However, both stories played a huge role in popularizing blockchain technology. The space is now maturing, both in terms of how it is perceived and in terms of regulations.

ICOs have been replaced with other formats for fundraising, which are more robust and often regulated. Such formats include launchpads, Initial Exchange Offerings (IEOs) on exchanges like Binance,[9] and Decentralized Exchange Offerings (DEOs) on decentralized exchanges like Uniswap.[10]

The last few years has also seen further developments both positive and negative, which have helped break the crypto world into the mainstream.

The community witnessed another "bull-run" as it's called, which saw the price of all cryptocurrencies spiking to new highs in 2022 before collapsing. The collapse was followed by a series of scandals, such as the complete collapse of the Terra/Luna stablecoin, the Genesis crypto-lending platform, and the bankruptcy of the FTX exchange, which was one of the largest at its peak.

At the same time, the world also witnessed new developments, such as decentralized finance, NFTs, and the metaverse.

Governments are discussing various crypto-regulations as well as rolling out CBDCs, which stands for centralized bank digital currencies, a centralized, digital version of money.

Therefore, while blockchain had a rocky start, it seems that it's here to stay and will play an even larger role in our lives in the years to come.

---

[8] www.icodata.io/stats/2018

[9] www.binance.com/en

[10] uniswap.org/

I Can't Trust You, But I Can Trust the Blockchain

"The root problem with conventional currency is all the trust that's required to make it work. The central bank must be trusted not to debase the currency, but the history of fiat currencies is full of breaches of that trust. Banks must be trusted to hold our money and transfer it electronically, but they lend it out in waves of credit bubbles with barely a fraction in reserve. We have to trust them with our privacy, trust them not to let identity thieves drain our accounts."

—Satoshi Nakamoto, Founder of Bitcoin

Blockchain is an interesting solution to a problem that many people wonder really exists. Transactions are ridden with issues of trust. Hence, any technology that promotes trust should be good for the economy. However, ICOs and the Silk Road demonstrated that some of the use cases promised by blockchains were not necessarily benevolent. Blockchain is supposed to solve issues of trust, but many ICOs were scams or simply bad investments, to the extent that there is a whole web page dedicated to this topic.[11]

So, how can blockchain solve the problem of trust, when we need to trust it first? Public blockchains have tried to solve a problem that is already solved through the global monetary system. One of the visions behind blockchain is that decentralization is good for its own sake, and that's not always the case. You don't always need to fix something if it isn't broken. Quite often, new technologies—blockchain included—are overhyped. But there are many enticing use cases for it.

First of all, there are arguments in favor of token economies in many places. For example, Brave, an open-source browser, created a token called the Basic Attention Token. This is a token that rewards users for watching ads, published for creating content on which ads are displayed (e.g., videos and blog posts). It completely removes intermediaries (like Google ads). IPFS (which stands for the Interplanetary File System) is a tokenized blockchain for sharing data. The users keep ownership of their data and can share it with each other, without having to go through a central server. Security tokens is another use case. Security tokens are cryptographic tokens that act as financial securities. Security tokens introduce all the benefits of blockchain into the world of traditional securities. For example, shares or land ownership titles can be

---

[11] deadcoins.com/

issued in security tokens. The benefits of doing that is that security tokens can be traded faster and 24/7 in contrast to traditional securities, for which trade has to go through intermediaries.

Another great example of an interesting token economy is Steemit.[12] Steemit is a social media website based on blockchain, with a very interesting token economy. There are three kinds of tokens: Steem, Steem Dollars, and Steem Power. The platform incentivizes content creators through rewards in Steem Dollars, which can then be converted to Steem, in order to exchange it with dollars, or to Steem Power. The latter represents ownership of the platform, and Steem Power holders receive tokens according to their shareholding power. Steemit demonstrates how a blockchain start-up can work in a democratic fashion. As blockchain is constantly evolving, there are many more examples that the informed reader can find.

Secondly, distributed ledgers have many applications in an enterprise setting. Corda, a distributed ledger technology (which is not based on blockchain), has been adopted by various companies as a way to streamline operations.[13] A consortium of banks used Corda to create a trade finance app,[14] because it can provide a better infrastructure than the existing one.

The use of blockchain in supply chains can reduce any information uncertainty and improve efficiency. Maersk found that a single shipment of refrigerated goods[15] involved 30 different individuals and organizations and 200 separate interactions. The complexity of this process is costly and prone to errors and fraud. Instead, blockchain can be used. The use of blockchain makes it easy to check the provenance of goods by any of the parties in the chain. A combination of Internet-of-Things and blockchain can be used to automatically track the location of goods, without the need of a human operator.

Big companies are already investing on that front. IBM is using blockchain as a solution for supply chain management through its TradeLens platform.[16] Walmart announced their intention to use the IBM Food Trust platform to trace food through blockchain.[17]

---

[12] steemit.com/

[13] www.corda.net/participate/index.html

[14] www.theglobaltreasurer.com/2017/08/08/banks-develop-trade-finance-app-on-r3-corda-dlt-platform/

[15] www-03.ibm.com/press/us/en/pressrelease/51712.wss

[16] www.tradelens.com/

[17] news.walmart.com/2018/09/24/in-wake-of-romaine-e-coli-scare-walmart-deploys-blockchain-to-track-leafy-greens

Blockchain also has multiple uses in the insurance industry. Etherisc[18] built an insurance program for flights. Insured passengers automatically receive compensation when a qualifying event takes place, which makes the whole process more efficient. They are now also working on crop insurance. EY has launched Insurwave,[19] a blockchain-based insurance platform for marine insurance. The Insurwave platform provides real-time information on ship location, condition, and safety conditions. When ships enter high-risk areas (e.g., war zones or storms), the program detects this and updates its underwriting and pricing calculations.

MIT has produced a blockchain-platform called MedRec for storing and handling medical records.[20] Patients can keep their records on the blockchain, and then they can choose to grant providers access. This lack of centralized intermediaries makes the system more secure and efficient, as the patients can have direct access to their records and full control of the way they are used.

Blockchain also has applications in government bureaucracy. Issuing ID cards and birth certificates, for example, can take place through blockchain. Taxation records can also be held on a blockchain system, and then automatically audited through smart contracts. The government of Thailand has put in place such a system for preventing tax-related fraud.[21] The World Identity Record[22] is a blockchain-based system that provides identity to those who do not have one or have lost it. People who have fallen victims of trafficking, victims of modern slavery, and immigrants fleeing from war zones are only some of the examples. The country of Moldova is one of the countries using this system in order to fight human trafficking.[23]

Blockchain companies are also trying to disrupt the financial sector, in things like remittance, money transfers, and tokenization of assets. Decentralized finance (DeFi as it is often called) has become a popular movement, with many groups working toward blockchain solutions to replace aspects of the current financial system. It's difficult to predict what the future of the financial sector will look like, but it will almost definitely include some blockchain applications.

---

[18] etherisc.com/

[19] insurwave.com/

[20] medrec.media.mit.edu/

[21] www.coindesk.com/thailand-government-trials-blockchain-in-fight-against-tax-fraud

[22] win.systems/

[23] un-blockchain.org/tag/moldova/

Finally, there are some other applications like the metaverse and non-fungible tokens (also called NFTs). The power of this narrative is such that Facebook decided to rename itself in Meta, in preparation for what might be the next big trend in computing. NFTs created a huge market in 2021, primarily focused around art. NFTs allow the exchange of ownership of digital assets (images being the best example), hence allowing the auction and acquisition of digital goods. Whereas in the past, digital goods (e.g., digital photographs) had very small value, because they could be replicated an infinite number of times, NFTs have turned these types of items into rare collectibles.

One of the main arguments against blockchain is that many of ways in which it is used can also be solved through a centralized database. It is true that in many cases blockchain is not needed. As happens all too often in technology, it has been overhyped and promoted as a cure-all. This trend is also encouraged by big companies, which are always keen to push new technologies to their clients. The arguments for and against blockchain are technical, but, at least for now, blockchain seems to be gaining ground in many industries.

If I had to summarize the core benefits of distributed ledgers, I would say the following. Distributed ledgers increase trust. The benefits of increasing trust can mean greater efficiency, greater profit margins, and the removal of intermediaries. There are technological challenges to blockchain, which we need to overcome if blockchain is to be adopted on a major scale.

Blockchain has the potential to play a huge role in the new economy of prediction that is emerging, which is the subject of the next chapter.

CHAPTER 15

Economies of Prediction: A New Industrial Revolution

> *"I visualize a time when we will be to robots what dogs are to humans, and I'm rooting for the machines."*
>
> —Claude Shannon

Ever since life started on Earth it faced the vague, the unpredictable, and the unknown. For the largest part of history, the unknowns of the natural world might have seemed more crucial than any human-made unknowns. Good or bad weather could spell the difference between starving or surviving for one more season. As societies increased in complexity, human-made unknowns became even more powerful than the natural ones. Wars eradicated millions of lives. New inventions led to unprecedented years of prosperity. New cultural and political trends emerge every few decades and change the way that humans live their lives and conduct business.

However, regardless of whether the greatest challenges were posed by nature or by other humans, throughout history, there has always been one theme in common: The desire of humans to predict and control the future. From the Delphi oracle of ancient Greece to the prophecies of Nostradamus, to shamans and mediums, humanity has always tried to predict the future. Uncertainty is a major cause of anxiety, one so strong that, as you read earlier, it is hardwired into our brains. The desire to avoid this anxiety is *so* strong that humans could easily be tricked by anyone claiming to possess mystical powers. Maybe the desire of many people to predict the future was such that they have even fooled themselves.

As humanity ventures into the 21st century, the ability to collect data, discover patterns, and make predictions is getting ever better and more accurate. The ability to predict the future is no longer science fiction; it has become a reality. We are witnessing a major revolution centered around data science. Some people are predicting that big data will completely revolutionize every aspect of our lives. Others are more cautious about the degree of disruption that will be achieved. In one way or another, however, artificial intelligence and data science are currently being employed by more and more industries.

Uncertainty plays a key role in many industries and it has even been the fundamental drive for some of them, like insurance. Hence, the ability to disperse the fog of uncertainty is of paramount importance. To understand the way in which uncertainty plays a role in economic activity, we have to categorize the different ways in which an industry might be exposed it. The three categories examined in this chapter are the following:

1) Uncertainty brokers: Industries whose core business is managing uncertainty and risk.

2) Incomplete information: Industries that are exposed to uncertainty about the true state of the world right now.

3) Prediction industries: Industries that are exposed to future uncertainty, and hence they have to predict future states of the world.

Another big impact in the economy takes place through automation. While automation is not directly related to uncertainty, the algorithms and technologies applied are similar. The categorization discussed here is based on economic terms.

The first category refers to all those industries in which prediction plays a key role. For example, in agriculture and fishing, the ability to predict the weather is of paramount importance, as bad weather could potentially spell disaster for a crop. But this concerns prediction about a natural phenomenon. Other industries are exposed to uncertainty about events influenced by humans. For example, in finance, if someone can build a great predictive algorithm for stock prices, wealth will automatically ensue. In the retail industry, companies have to predict the future demand for products, as well as new trends. In policing, the authorities could benefit from being able to predict the spots where new crimes are most likely to occur. And these are only a few of the different examples of industries exposed to future predictions.

The second category refers to industries that are either exposed to unknown cause-and-effect relationships or that need to make decisions based on incomplete information. For example, in medicine, it is quite often difficult to predict how a medicine will affect an individual. A doctor has to make an educated guess about the side-effects or the potential effectiveness of the drug. Netflix is trying to predict which movies you are going to like the most, and Amazon is predicting the products you are most likely to buy. The military has to make decisions based on incomplete information about an enemy's capabilities. Judges have to make decisions based on results from forensics.

Finally, there are some industries where uncertainty is the product that is being traded. The insurance industry is the prime example of this. Since its inception, it has provided assurances against risk to those who find the risk detrimental to their businesses and lives. The finance sector also belongs in this category. Many financial products attempt to ensure profits while minimizing risk. People investing in those products are looking for a safe haven for their money, in order to protect against inflation, rather than wanting to get rich by predicting the stock market.

Uncertainty Brokers

If I have to choose one industry that is mostly related and exposed to uncertainty, it has to be the insurance industry. Insurance has a long history, with first records appearing in Babylon as early as 4000-3000 BC.[1] Babylonians practiced *bottomry* contracts, according to which, a merchant who was taking a loan didn't have to repay the loan if the shipment was lost at sea. Those types of contracts were also practiced by Hindus, ancient Greeks, and Romans. The ancient Romans also had "benevolent societies" that pooled money to take care of funeral costs for their members.

With the importance of marine trade since ancient times, it makes sense that the first types of insurance involved ships. The first actual insurance contract for maritime insurance dates from Genoa in 1347.[2] Fire insurance was developed much later. After the great fire of London in 1666, the need for fire insurance became apparent, with many companies popping up. In the United States, the first insurance company was organized by Benjamin Franklin in 1752, and the early 20th century saw the rapid development of life insurance.

---

[1] www.britannica.com/topic/insurance/Historical-development-of-insurance
[2] Franklin, James (2001). *The Science of Conjecture: Evidence and Probability Before Pascal.*

One of the most important historical developments in insurance was Lloyds of London, which became the first international insurance market. What many people don't know are the unusual beginnings of the company. Edward Lloyd was the owner of a coffee house that opened in 1687 near the Thames on Tower Street. It was a favorite of people related to shipping, so the coffee house was always up-to-date with current news. In 1696, they published Lloyd's list, which contained information about ship arrivals and departures, as well as sea conditions. Captains would compare and improve routes with each other in Lloyd's café. The list eventually grew to a publication that included information on everything from foreign markets to high-water times at London Bridge to accidents.

Lloyd's eventually started attracting people who were willing to insure ship owners and captains against risk. Those people would seal the deal by writing their name under the terms of a contract. Hence the term "underwriting" was born. While insurers operated as free agents for a long time, they grouped together in 1771, creating the Society of Lloyd's.

The importance of the insurance industry can easily be demonstrated by the work of Kenneth Arrow (1921–2017). Kenneth Arrow was one of the most prominent economists of the past century, and he was awarded the Nobel prize in economics in 1972. Arrow was one of the first economists (alongside Keynes and Frank Knight) to focus on the economics of information, that is, how uncertainty plays a role in the economy. The Arrow-Debreu model used the concept of a "complete market," that is a market where it is possible to trade on every possible state in nature, and information is complete (e.g., the seller does not possess more information about a good than the buyer). The model differentiated the concept of a commodity by space and time.

For example, apples in New York in September are a different commodity than apples in London in February. Such a market can reach a Pareto optimal equilibrium, which, in everyday terms, is the goal of an economy that wants to maximize prosperity. Without the insurance industry, it is impossible to do that, which is why Arrow applauded insurance and other risk-managed tools like futures contracts. The market for risk makes the world go around, as it allows entrepreneurs and business to take risks that they otherwise wouldn't take. Insurance even has its own insurance industry, called re-insurance, which balances out some of the risk that insurers take.

In the category of uncertainty brokers we could also include low-risk funds. The main idea of these funds being that they can protect your savings, but they are not wealth-generating machines. However, regardless of the risk that a fund uptakes, the rules of the game mean that the fund manager still has to predict the market to one extent or the other and hedge the risk. Futures contract are also a good example of this category.

A futures contract is a legal agreement to buy or sell a particular commodity at a particular price at a particular point in the future. A farmer (who is a seller)—who is worried that the crop might not yield enough food because of bad weather—can sign a future contract, which ensures that the food can be sold at a particular price, ensuring the viability of the business. A buyer—who is operating on short margins and is particularly worried about the price of a commodity, like gold—can agree to buy it at a particular price, making it easier to plan for the future.

In funds and futures contracts, the nature of prediction is somewhat different than with insurance. Insurers are predicting probabilities of events that will more or less be stable over a large period of time. Let's say that you run your own insurance company, and you are insuring against cardiovascular disease. The risk of cardiovascular disease in a country is relatively stable over a long period of time. We wouldn't expect that risk to change drastically in one year, for example. It would probably take changes in lifestyle or nutrition to dramatically change this risk up or down. There are some other risks related to natural disasters or terrorism, which might be more unpredictable, but still the same principle applies. In most cases, the risks are stable over long periods of time. The main issue that insurers facing is getting access to information that helps them calculate the true probabilities of events.

However, in banking and finance, interest rates and prices can fluctuate greatly over the course of a year. Hence, the ability of those industries to survive also depends on their ability not only to hedge risk, but also make predictions about them. Even for safe non-speculative financial instruments (such as Swiss bonds), the purchaser is still making a prediction about the stability of the economy in some future point in time. Hence, the nature of funds might place them more in the category of "prediction industries," even if many investors are using funds as a way to guard against economic uncertainty and inflation.

So, how is data science changing the business of uncertainty brokerage? Uncertainty brokers care about two things: the probability of an event and the magnitude of an event. An insurer runs a business similar to a casino. In a casino, those who lose money pay those who gain money. If there are more people who lose money than who gain money, the casino will be profitable.

A casino has the luxury to amend the rules of the game in a way that it will always be profitable. However, an insurer has to predict the probability of events. And this is where data science is stirring things up. As collecting data becomes easier and easier, and predictive algorithms become more and more powerful, predictions, inevitably, become more accurate.

Let's examine health insurance. Using fitness trackers, it's possible to get a good idea of metrics like someone's resting heart rate and their average activity levels. Using this data, the insurer can get a much more accurate idea of someone's health profile, and the risk premium can be adjusted dynamically. Hence, as information improves, the insurance market reaches an improved equilibrium. Those who are healthier pay smaller premiums, compared to those who are higher at risk. A similar approach has been implemented in car insurance.[3] You can install a monitoring device in your car, and if you are a safe driver, you get a discount.

These approaches might raise various moral and practical questions. First of all, the soundness of the plan assumes that the fitness tracker can't be gamed and is accurate. There is the moral hazard of people tricking fitness trackers into recording more steps than they actually do, for example. The second question is how insurance would deal with those cases of people who are at risk, but there is nothing they can do about (e.g., they suffer from some genetic disorder). Another issue is privacy. Are we willing to reveal our personal data?

Regarding the first point, while it might be possible to game fitness trackers, there are various ways to get around this problem. First of all, fitness trackers improve all the time. Secondly, health insurance companies might be using time horizons of many years. So, even if someone has found a way to game the tracker, they might not be willing to do this over a span of ten years. Thirdly, even if some people *do* get around the tracker, it is unlikely that everyone will be doing it. Finally, fitness trackers are not the only way to record someone's fitness levels. Attendance to gym sessions can be another one, for example—although again there's no guarantee that actual effort is exerted, even though the person might be physically present.

One of the questions about this new trend is how far insurance companies should go into handling our personal data. Many people might just be unwilling to share their personal data with companies. It's likely that there is no good answer to this question, and this is still a hotly debated topic, not only about insurance, but about any industry relating to the Internet. The business model of many big companies, like Google, has always been based on this motto: "Use the service for free. We get your data and we sell it." A series of recent scandals, like Cambridge Analytica, made people more aware of how their data can be used in the wrong way. Additionally, the European Union's General Data Protection Regulation (GDPR) restricts the collection and use of data.

---

[3]`www.allstate.com/tr/car-insurance/telematics-device.aspx`

Regulations won't necessarily solve the problem. It is inevitable that some business models depend on data availability, and more information can increase the efficiency of a market. Hence, as more consumers decide to enjoy better premiums through tracking, other consumers will inevitably be left with a tough choice. Education and awareness about the benefits and challenges of sharing private data are even more important than regulations.

One of the benefits of recent technological developments is that sharing private data does not mean that privacy is not respected. Privacy-ensuring data analysis can be achieved through various solutions. First, analyzing data does not require the involvement of a human. There is a huge trend right now around automated machine learning (autoML). An algorithm finds a good machine learning solution without intervention from a human. An insurer can store data in a central repository, which is then automatically analyzed by an autoML solution.

Another interesting approach is homomorphic encryption. This is a type of encryption that allows the analysis of encrypted data without decrypting it. Hence, the data can be used without anyone being able to access it.

Another important solution in this space is blockchain. Blockchain increases trust in transactions and the recording of data. Combining blockchain with AI can be a solution for dealing with claims, fraud, and privacy in one go.

Like Arrow predicted, the natural outcome of a market where information becomes more complete is that it becomes more efficient. But what happens when someone suffers from a condition that is incurable or genetic, and they can't do anything about it? While data might predict that these customers are going to be a risk for an insurance company to insure, it is data that will make protecting those citizens easier. The question of how to assist people with disabilities or genetic disorders is not necessarily a technological concern, but more of a political and moral one. In various countries, like the United States and the UK, there is a heated debate about the current state of the health system, and the best balance between a public and a private healthcare system.

However, in one way or another, making markets more efficient means more wealth will be generated, which increases available funds to spend on less fortunate members of society. Kenneth Arrow observed[4] in 1978 that:

> "The special economic problems of medical care can be explained
> by adaptations to uncertainty in the incidence of disease and in
> the efficacy of treatment."

---

[4] Arrow, K. J. (1978). "Uncertainty and the Welfare Economics of Medical Care. In Uncertainty in Economics" (pp. 345-375). *Academic Press.*

Uncertainty plays a central role in healthcare, and any technology that helps dissolve uncertainty will help improve healthcare.

We've categorized insurance companies as uncertainty brokers, but they also utilize prediction and operate under incomplete information. The main distinction of uncertainty brokers from the other industries is that uncertainty brokers trade uncertainty directly. Let's look at some industries in which incomplete information is a major pain.

Industries of Incomplete Information

Chances are you have used an e-shop like Amazon or a service like YouTube at least once in your life. When you use a service of that kind, you will notice that the website recommends products or content to you. Quite often you find those products or content recommended to be quite close to your interests, which might make you consume even more content and buy more products. There is an algorithm running in the background, reading your past purchasing behavior and making sure that it recommends products and content you are going to like. Those systems are called "recommender systems" and they are a huge part of the online retail space.

Until online shopping became a thing, retailers had to make assumptions about what their customers would like. The merchant is in a constant state of uncertainty. In some industries, like fashion, the trends change every year. In some others, like electric appliances, this happens less often. But in one way or another, all merchants face the question of what is a customer going to like. They operate under incomplete information.

Data science shaped the way of all that. Amazon stores your complete purchase and search history. Facebook holds your full history of likes, shares, comments, and interactions with other people. Google knows every search you've ever done. The tech giants might have a better understanding of what you will like before even you know you will like it. They are feeding this information into machine learning models, in order to keep you engaged and paying.

Note that the benefits of technology are asymmetrical. Industries that are employing data science to sell more goods and services are increasing their profits. The benefit to the end user is using a good or a service that might have otherwise gone unnoticed. However, this assumes that the user is better off using the aforementioned service or consuming the aforementioned good. There is lots of conversation around the impact that social media can have on our health, and part of the problem is the fact that algorithms have become so good in predicting what we'll like.

Incomplete information shows up in other contexts, like market research. When a company plans a new product launch, it has to assume that the reception is going to be positive. But it is not possible to be totally certain of that. Hence, any tool or method that can remove uncertainty is very valuable. Netflix's case study of how it got into the French market is a great example of how data science can be utilized in this context. Netflix used topic modeling to analyze two years of digital conversations to get the pulse of the French population.[5] A *topic model* is a natural processing technique used to extract the topics of conversation from large volumes of text. Hence, huge datasets of conversations can be summarized in a few sentences. Netflix used the insights from this research to launch a smart promotional campaign that included offers and ads that resonated with the French audience. At the time of writing, there are more than five million Netflix subscribers in France.[6]

Netflix was revolutionary in its use of data in other ways. It used data in its product creation process. The successful show *House of Cards* was based on data analysis. Netflix discovered that the director David Finch was very popular. They also discovered that viewers who liked David Finch also liked the 90s British political drama *House of Cards*, as well as films starring Kevin Spacey (although likely no longer). Netflix bought the rights to the series to create an American version of it. Then it used different marketing strategies for different types of users. The show ended up landing Netflix 3 million new subscribers.[7]

Personalization is the name of the game in those industries. These days, the wealth of user data and the availability of powerful algorithms and computing power allow companies to come up with new products and strategies tailored to each user, as Netflix demonstrated.

Besides the consumer industry, this also has implications for medicine. A new trend in the medical sector is called precision medicine. Read any medical study, and in most cases the results will not be entirely clear. The study will report that 20 percent of the patients developed symptom A as a side-effect to a drug, and 30 percent developed symptom B. It is difficult to predict *a priori* how patients will respond to drugs, or how

---

[5] www.youtube.com/watch?v=tZkILxaANLU
[6] www.statista.com/statistics/607819/netflix-subscribers-in-france/
[7] www.youtube.com/watch?v=OFAqbfEl870

diseases will progress. Precision medicine personalizes the treatment according to each patient's genes, environment, and lifestyle. The technology is progressing to the point where it will be accessible to a large number of people in the future, with UK's NHS[8] and UCL[9] investing in this trend.

AI has also progressed greatly in the creative domain in tasks like image generation and music creation. Google "generative adversarial networks images" and see some of the images that neural networks have created. The technology has reached a point where generated human faces cannot be easily discriminated from fake ones. The pop star Taryn Southern created an album called *I Am AI*, which was generated by AI.[10]

It might be possible in the near future to get unique pieces of music or images that completely resonate with someone's tastes. Whereas artists and companies had to predict the pulse of society in previous times, now this information can be mined, and it can also be used to tailor products to match each individual's tastes.

Prediction Industries and Automation

Some of the industries that are also being affected by data science are the ones in which prediction plays a key role. Finance and investing being prime examples, but not the only ones. The whole hedge fund industry has been on prediction, with quants playing a huge role in it, and the majority of hedge funds utilizing AI in one way or another.[11]

Finance is probably the prime example of an industry that is an uncertainty broker and a prediction industry. Derivatives are a great example of this. Futures contracts are used to buy and sell a good in the future at a predetermined price. Options is another type of derivative that give the party the ability to buy or sell to the other side at a predetermined price in the future. Prediction plays a key role in both cases. The industry is waking up to the predictive power of machine learning algorithms,[12] with automated trading playing a larger role every year.

---

[8] www.ucl.ac.uk/precision-medicine

[9] www.ucl.ac.uk/precision-medicine

[10] www.digitaltrends.com/music/artificial-intelligence-taryn-southern-album-interview/

[11] www.barclayhedge.com/majority-of-hedge-fund-pros-use-ai-machine-learning-in-investment-strategies/

[12] www.ft.com/content/8cc7f5d4-59ca-11e8-b8b2-d6ceb45fa9d0

Agriculture is another industry where predictions play a huge role, with the weather having a huge effect on farmers. Machine learning can also be used to discover the optimal conditions for plant growth and boost yields. Computer vision can be used to detect early signs of disease.[13] Automation (which I come back to shortly) can also help reduce the costs of production.

Urban data is also another important area. Cities are complex organisms that face complex issues like traffic and pollution. Data science can be used to predict bottlenecks before they take place and discover the best routes for buses. Uber, which is one of the most data-driven companies in the world, is investing in urban data.[14] Uber would never have existed was it not for the power of data science and prediction. It is constantly optimizing routes and prices, in order to match drivers to riders and destinations.

The increase in computational power and improvement of predictive algorithms means that any industry in which prediction plays a role sees improvements in efficiency and profits. The degree and the extent to which this happens depends on the specifics of each industry. However, the constant improvement of data science products means that this technology is also accessible to smaller enterprises, and not only big corporations. Even small benefits can diffuse across the whole of economy.

No conversation about the impact of data science in economy would be complete without talking about automation, which is closely linked to the problem of prediction. While I chose to categorize industries in this chapter in three distinct categories, the distinction is vaguer from a technological point of view, with common algorithms being employed in every case.

Automation is being employed more and more, with some companies owning part of their success to it. For example, an article in the *Economist* explained how Amazon is investing heavily in automation in its warehouses, thus being able to reduce costs and delivery times.[15] In the near future, a large number of supermarkets will not have many human employees, as machine learning will be employed to make payments automatic, without someone having to go through a cashier point.

---

[13] www.technologyreview.com/s/612056/how-machine-learning-and-sensors-are-helping-farmers-boost-yields/

[14] movement.uber.com/

[15] www.economist.com/business/2019/04/13/amazons-empire-rests-on-its-low-key-approach-to-ai

It is believed that many blue-collar jobs, from drivers to construction, are threatened by automation. This is true to a large extent. However, what is also true is that some white collar jobs are threatened too. For example, AI is used in the legal sector to discover relevant legal material and check contracts.

There is widespread fear about the impact that automation could have on the economy. Some people are arguing that many jobs will be lost, and the only solution is to have a universal basic income. Some others are more optimistic, saying that humans will have more time for creative endeavors. Others, using examples from the past, say that while every new technology destroys many jobs, it also creates new jobs.

If this book has taught you anything, is that it is not easy to predict the future by using only data from the past. The truth is that we can't really know the effects of automation. The effects of data science in the three types of industries (uncertainty brokers, industries of incomplete information, and prediction industries) are intended to incrementally improve efficiency, reduce costs, and improve profits.

Automation, on the other hand, is massively disruptive. It completely changes the nature of some industries. Hence, we might be facing an unprecedented non-linearity of the kind that complexity theorists have been studying. Whether this will be good or bad, it is impossible to know. Like it always happens with complex systems, once an innovation emerges in a system where competition and survival are the two key attributes, it is impossible to go back.

Probably the best example of this is ChatGPT and Large Language Models, which made a huge impression on the public in late 2022. ChatGPT, along with other language models, represents a significant breakthrough in natural language processing (NLP) and artificial intelligence (AI). By training on vast amounts of text data, these models can generate human-like responses to a wide range of queries and prompts.

This means that many white collar jobs, such as copywriting, can be largely automated using this technology. However, no matter how impressive this technology is, it still does not constitute true AI. However, it demonstrates the power of deep learning and offers us a very good glimpse into what the future will look like.

The Global Economy Against Uncertainty

There are many over-optimistic proponents of data science who believe that we will soon be able to predict everything. From new fashion trends to diseases, our ability to predict becomes better and better. Hence, someone might extrapolate that uncertainty will eventually be reduced to zero.

Then again, there are some more pessimistic voices. Nassim Nicolas Taleb has expressed the opinion that we simply can't predict significant future events. There are two main reasons for that. First of all, our ability to model things depends on the past occurrences. The biggest flood or disaster might have simply not happened yet. Secondly, a large part of the uncertainty in our current world does not come from the natural world, but from the social world. Predicting financial crises is even more difficult than predicting natural disasters. Hence, it is futile to try to make predictions. Rather, we have to try to build *antifragile systems,* as Taleb calls them. These are systems that benefit from randomness and uncertainty.

For example, a big corporation with multiple layers of management and strict processes is fragile. Any radical change in the economy can quickly turn this organization into dust. On the other hand, a small technology company can be very antifragile. A single random event (e.g., signing up a big client) can quickly turn this start-up from a small company to a rising star.

A complete discussion of the concept of antifragility is beyond the scope of this book; however, it holds some merit. Nevertheless, there are some points in both viewpoints, and, as always, the truth is more complicated.

For many real-world challenges, we might eventually be able to have near-perfect predictions, especially with anything relating to natural systems. Medicine is a good example. Also, social systems with bounded rules can be predicted with a high degree of accuracy. Traffic jams are a good example of that, as agent-based modeling can be used to optimize traffic.

There are some natural phenomena, like floods and earthquakes, which might be difficult—if not impossible—to predict over a long horizon. These are natural, complex systems, and much like complex social systems (e.g., an economy), a variety of factors combine to make prediction nearly impossible. However, we saw how simulations can be an invaluable tool in those cases.

My personal opinion is that discussing whether we will be able to perfectly predict everything far into the future brings us back to the root of the big questions around uncertainty. Is all uncertainty in the world aleatoric or epistemic? Can we, like the Victorian scientists believed, predict everything as long as we can measure it, or do we live in a chaotic world, where our ability to predict uncertainty is bounded?

These are important philosophical questions, but doing is usually more powerful than just thinking. And while we have been pondering these questions, economies and societies have faced improvements in efficiency and profit. These improvements, however, have been gradual and incremental. We hear voices warning us of some impeding singularity, where AI will take all our jobs, and the world will come to a standstill. The truth is that automation *will* claim many jobs, but we are not risking a worldwide AI takeover.

Our ability to make predictions becomes better and better. Some people can actually make good predictions, even if they are in the minority. Our brains have developed to deal with uncertainty, and the fact that we are very much alive now demonstrates that at least some of the uncertainty in our world can be defeated.

However, as we progress, there are new uncertainties included in our world. We don't have to worry where our food will come from, but we have to worry about the interest rates set by the central bank and how they will affect our mortgage, or about the result of the next election. We have become great at removing uncertainty in many small things, and this came as the result of complex interconnected societies, where this kind of data aggregation is possible.

However, in doing so, we have increased the complexity and unpredictability of other aspects of our society. On an economic level, a crisis in one country can bring the global financial system down. As we conquer the predictable uncertainty of the natural world, humans become the main source of uncertainty. It is up to us to control those parts of social systems that can reduce uncertainty. A financial system doesn't have to be fragile. It was the designer's mistakes that made it so. Governments can be a great source of uncertainty, but it is within our power to use technologies to make decision making less centralized and better organized. Guided by the right principles, uncertainty can be reduced in any social system. However, this goes beyond the domain of simple technological progress, and into the domain of social progress, which is always slower and has many ups-and-downs. Then again, a bit of uncertainty makes our life more interesting, so we might never want to get rid of it.

EPILOGUE

The Certainty of Uncertainty

In this journey, you saw various heroes. Ancient philosophers like Aristotle touched upon the problem of uncertainty, which their more modern counterparts, like Hume, tried to answer. The mathematicians of the Renaissance came up with the first quantification of uncertainty in the form of probability theory, which formed the basis of the discipline of statistics. The Enlightenment brought new concepts and tools like the law of large numbers and the central limit theorem, proving that even in the face of uncertainty, nature seems to follow certain regularities.

Fisher gave us the foundations of statistics, and the methods in research design that are still being used in practice, while hypothesis testing is currently ridden in controversy. Bayesian statisticians had operated in parallel to the traditional statistical establishment, until recently where Bayesian methods, invigorated through the Monte Carlo methods and the increase in computational power, have become mainstream.

Shannon was the first person to quantify uncertainty, and machine learning has made heavy use of his theories. Kahneman and Tversky showed us how are minds respond to decision-making under uncertainty through various biases. Neuroscience is studying the way our brain responds to uncertainty, with Karl Friston arguing that uncertainty is at the core of all cognition.

Currently, "data science" is one of the hottest buzzwords in technology, merging machine learning, artificial intelligence, statistics, and all other data-related disciplines. Blockchain is another hot topic right now, promising greater efficiency, more transparency, and less uncertainty about the content, trust, and nature of transactions.

© Stylianos Kampakis 2023
S. Kampakis, *Predicting the Unknown*, https://doi.org/10.1007/978-1-4842-9505-2

Judea Pearl rose awareness of the issue of causality in our current models, arguing that we only see a small part of the whole picture. Complexity theorists have tried to pierce the layers of complicated interactions between interconnected parts and explain how the world works, and you learned how some of their methods faltered, but others are still driving innovative research.

I also talked about forecasting and the challenges of predicting the future. And I talked about the limits of prediction and whether nature imposes some ceiling on our ability to remove uncertainty. While there have been radically different approaches to this topic, the truth is that no one has the definite answer yet and we might never really know. While we are searching for this ultimate algorithm, method, or predictive machine, the economy and society around us are changing at a rapid rate, as the result of the very algorithms developed to fight uncertainty.

Improved efficiency, automation, better predictions, distributed ledgers—the promise of the new world dawning in the 21st century is that of a world where uncertainty is constantly being reduced. Whether we will ever be able to completely remove it from our lives, or whether this is even necessary, makes for an interesting debate. Probably, most of us would rather get away with the unnecessary uncertainties of life, those that cause us anxiety and stress. However, many would argue that a bit of uncertainty makes our lives more interesting. What's life without a surprise or two?

What we do know is that uncertainty has played a huge role in the evolution of our species, starting from our brains, and moving on to the formation of our societies, and the development of science and technology. It is time that we give it the respect it deserves, but also recognize that we don't have to be afraid of it. Humanity has survived uncertainty through millions of years of evolution, and individuals have thrived in the face of it. It can be a powerful motivator, a force of change, and, in some cases, a force of progress. As technological progress adds more layers of complexity and interconnectivity to our lives, everyone tries to predict what the implications will be, how the world will look, which jobs will be destroyed, which jobs will be created, and so forth. If this book taught you anything, it is that you can never be completely certain about how the future will look. Part of this book was written and edited during the COVID-19 pandemic, an event that very few people expected. This crisis will potentially have huge social repercussions, and at the time of writing the Internet is full of articles trying to predict the long-term effects.

However, much like in the whole of human history, the future belongs to those who can face uncertainty, overcome it, control it, and benefit from it. Learning how to do this in the best way possible and developing the tools to do so, is a worthy endeavor. The societies, businesses, and individuals that embrace this struggle are the ones that are going to thrive.

And of that, I am (fairly) certain about.

Index

A

Agent-based modelling, 185–188, 207–209
Aleatoric uncertainty, 5, 6, 52, 54, 105, 166, 207
AlphaGo, 120
AI winters, 116
Analytic concepts, 140
ANOVA test, 100
Antifragile systems, 251
ARIMAX, 152, 153
Arrow-Debreu model, 242
Arrow, K., 242
Artificial intelligence (AI), 131, 250
Artificial neural networks (ANN), 127
Astronomy, 42
Automated machine learning (autoML), 245
Automation, 249, 250, 254
Autoregressive Integrated Moving Average (ARIMA), 152, 155, 157, 158
Autoregressive process, 154
Axtell, R., 187

B

"Ban", 47
Battleships board game, 41
Bayesian decision theory, 48
Bayesian inference, 31
Bayesian interpretation, 162
Bayesianism, 55, 162
Bayesian networks, 45, 46

Bayes, T., 17
Bayes' theory, 41, 43
Belief Functions and Applications Society (BFAS), 74
Bem, D., 96
Bernoulli, J., 42, 61, 203
Beta regression, 107
Bias-variance trade-off, 121
Binomial distribution, 57
Birthweight paradox, 142, 144
"Bit", 47
Bitcoin, 159, 170, 171, 227–229, 231, 232
Black-Scholes model, 159
Black swans, 25–27
Blockchain, 160, 228, 231, 253
Boltzmann, L.E., 180
Boltzmann machine, 181
Boole, G., 73
Bottom-up approach, 116
Box, G., 99
Brexit, 161
Brown's double exponential smoothing, 155

C

Cardano, G., 61, 178, 203
Casinos, 178
Causality, 134, 137, 139
  association, 144
  causes, 139
  counterfactuals, 144
  intervention, 144

Printed in the United States
by Baker & Taylor Publisher Services